Introducing
Biochemistry

E. J. Wood

Department of Biochemistry
University of Leeds

W. R. Pickering

Senior Science Master
The King's School, Canterbury

John Murray

© E. J. Wood and W. R. Pickering 1982

First published 1982
by John Murray (Publishers) Ltd
50 Albemarle Street, London W1X 4BD

Reprinted (revised) 1984, 1987, 1989

Printed in Great Britain by The Alden Press, Ltd

British Library Cataloguing in Publication Data
Wood, E. J.
 Introducing biochemistry
 1 Biological chemistry
 I. Title   II. Pickering, W. R.
 574.19'2   QH345

ISBN 0–7195–3897–1

Foreword

Traditionally, biochemistry has been a subject that is usually taught at University level. During the past two or three decades, however, the tremendous advances in understanding the chemistry of the cell have had a profound effect on the teaching of biology, and to a lesser degree, chemistry in the schools. Inevitably this has meant that more and more material that was originally part of first year University courses in biochemistry has been introduced into A-level syllabuses. The amount of biochemistry taught has varied from school to school and has been a reflection of the nature of the A-level syllabus and the background of the teacher, only the younger of whom have taken courses in biochemistry at university. Due to the shortage of good elementary text books of biochemistry suitable for school use, students have resorted to consulting the plethora of textbooks intended for the university student. Excellent though many of these books are, they are not ideal for the A-level student of biology and chemistry. They are even less suitable for the serious student of any age who seeks to understand something of the importance of biochemstry in the modern world.

The Biochemical Society, the professional body in the United Kingdom concerned with the advancement of biochemistry, has been very conscious for some years of the difficulties in providing adequate instruction in biochemistry at the school level. With the aim of resolving this problem the Society has encouraged Drs Wood and Pickering to write a textbook of biochemistry for schools. The results of their efforts, 'Introducing Biochemistry' describes in simple language, using the minimum of chemical formulae, what biochemistry is about, how it explains the complex workings of the cell and the impact of this knowledge on our everyday life. It is particularly appropriate that a simple book of this kind should appear today when we stand at the beginning of the biotechnological revolution. This revolution is taking place because man has now learnt to manipulate the complex chemistry of the cell. Some knowledge of this subject is therefore essential for the intelligent layman who wishes to appreciate the tremendous potential of the new technology.

The book has been written primarily for sixth-formers, but many other students taking courses with a biochemistry component, ONC, Nursing, Environmental Science, etc., will benefit from the broadly-based text, indicating clearly the importance of biochemistry in all living processes. It is hoped that it will become the standard biochemical textbook for students intending to study biochemistry or the biological sciences at Universities or Polytechnics and that it will ensure that they are able to start undergraduate work with a similar basic knowledge of the subject.

S V Perry
Professor of Biochemistry. University of Birmingham.
Chairman of the Committee of the Biochemical Society.   October 1982

Preface

Living organisms are immensely complicated, and the Biochemists aim is to explain, in chemical terms, how living organisms work. Although Biochemistry is quite a young science compared with, say, Chemistry and Physics, it has been remarkably successful, especially in recent years, in achieving this aim. Because of this success, Biochemistry nowadays forms a fundamental part of the 'life sciences' and has strong connections with Biology, Medicine, Agriculture as well as with industries such as food-processing and pharmaceuticals. People being trained for work in these areas need to have a knowledge of Biochemistry, and therefore most GCE Advanced Level Biology syllabuses today have in them a considerable biochemical element. Biochemically-biased questions occur in Scottish Higher Biology Papers and Biochemistry options are also offered in some Advanced Level Chemistry syllabuses. Other courses such as T.E.C., National Certificates and Diplomas in Biology, as well as Nursing Studies, require Biochemistry as an integral part of the course.

This book was written with the people taking such courses and examinations in mind, and its aim is to show that Biochemistry is not difficult to *understand*, and may actually be enjoyable. There is, it is true, quite a lot to *remember*—chemical structures, names of enzymes, pathways of metabolism, and so on—but we believe it becomes easier to remember these if the student can understand the principles of how living things operate. The principles themselves are quite simple, and also they are, on the whole, the same in a human cell, a plant cell, or a bacterial cell.

In this book we have tried to emphasize the 'simple principles of life chemistry' at the expense of chemical detail. We recognize that there is a danger of oversimplification in this approach. However we think it better to accept this risk and keep the student interested and encouraged rather than insist on his learning of a multitude of facts and chemical structures that seem to be unconnected and leading nowhere. For these reasons the chemical details of the structures of the 'molecules of life' are not emphasized as much as they might be in a traditional textbook of biochemistry. This also recognizes that there will be some taking 'A' level Biology who are not simultaneously doing 'A' level Chemistry. Structures are frequently drawn as outlines or as simple geometrical shapes, especially early on in the book. By doing this we hope we can, even for those with minimal chemical knowledge, explain some of the principles of 'life chemistry'. For example *catabolism* means chopping big molecules up into smaller ones with the release of usable energy: molecules 'recognize' one another because their three-dimensional shapes 'match', and so on. This is not to say that Chemistry is unimportant. Simply, having understood the *reason*, we then have the incentive to go on to see *which* particular chemical substances perform particular functions and *how* they do this.

The book is set out in five large sections or chapters, each with a broad theme, but there is a final sixth chapter which deals with the impact of Biochemistry on life today. The first five sections deal with the structure of cells and of the 'molecules of life', especially macromolecules, ways of obtaining energy and of using it to support life, and lastly the ideas about information and biochemistry, an area which is very closely related to the science of Genetics. Many of the study questions at the ends of chapters are taken from GCE Advanced Level papers.

It is our belief that complicated and delicately-balanced living systems can be understood even without much basic knowledge of chemistry, and we ask 'Can an educated person, in the 1980s, whatever his final choice of career, afford *not* to understand the significance of the impact of Biochemistry on our lives in the last two decades of the Twentieth Century?'

Contents

Acknowledgements

Professor P N Campbell, Director of the Courtauld Institute of Biochemistry, London, instigated the writing of this book and the authors are grateful to him for constant encouragement. The Biochemical Society has, through its various Officers and Commit- tees, supported the project from its inception, and the authors would like to record their thanks for this support and also to acknowledge the help given by various Officers of the Society in numerous ways. Many other individuals contributed along the way by reading all or part of the manuscript and making helpful comments and suggestions. We would especially mention Professor Ramsey Bronk, Miss Grace Monger, Dr Michael Roberts and Dr Barbara Gray and we are extremely grateful to them. The manuscript is surely clearer and easier to read because of their efforts. Any errors or omissions, in contrast, are our own and we hope that readers will point them out to us.

Many of the electron-micrographs are the work of Douglas Kershaw who gave freely of his time and expertise, and the electron-micrographs themselves represent 'snapshots' along the way of various research projects going on in the Department of Biochemistry in the University of Leeds. We are grateful to others who have generously provided pictures and photographs and their individual contributions are acknow- ledged in the figure legends.

Many of the study questions printed at the ends of Chapters are taken from recent GCE Advanced level or Scottish Higher Biology papers, and we record here our grateful thanks to the following Boards for allowing their questions to be used: Joint Matriculation Board, Oxford and Cambridge Schools Examination Board, The Associ- ated Examining Board and the Scottish Certificate of Education. Examining Board. Questions originating from the papers set by these Boards are indicated by [JMB], [O&C], [AEB] and [Scottish Higher].

We are extremely grateful to Mrs Sue Sparrow for translating our ideas into diagrams and to Mrs Sandra Gray for typing the manuscript. We are indebted to Harvey Johnson, of John Murray, for the sympathetic way in which the manuscript was treated and for constant encouragement, and to Debra Myson-Etherington for editing it skilfully and speeding it to the press. Last, but by no means least, we acknowledge the support given throughout by our respective wives in all sorts of ways too numerous to mention but deeply appreciated nonetheless.

1
Biochemistry and the Ecosphere

Summary

The Science of Biochemistry aims to describe the structure and activities of living organisms in chemical terms. Chapter 1 sets the scene and describes the players in the game of life. We see which of the elements of the periodic table are used by living systems and what compounds these form. Because the chemical changes in living organisms take place in an aqueous environment water has a key role: life is only possible because of the special properties of water. Practically all life on earth depends on sunlight as its energy supply for continued existence. Plants trap this directly: animals eat plants or other animals, and most of the elements of life are recycled. The many chemical reactions going on in organisms need to be properly organized and kept separate from the environment. It is the function of membranes to perform such tasks. Cells are not mere 'blobs of protoplasm' but are highly structured. Eukaryotic cells particularly are characterized by their possession of subcellular organelles such as mitochondria and chloroplasts which have special roles to play in the cell.

1.1 Introduction

Biochemistry is sometimes described as the chemistry of life. It is a relatively new scientific discipline and uses ideas derived from chemistry, physics and biology. Biochemical research is a very important part of all biological research today and has had a profound influence in many areas including medicine, agriculture and the food industry (Figure 1.1). In order to understand what biochemistry is all about it may be helpful if we start by taking a brief look at how biochemistry evolved.

What are the origins of biochemistry?

Chemistry developed as a science in the latter part of the eighteenth century. As knowledge increased, people began to ask what distinguished living matter from non-living matter. Both of these were clearly made up from chemical compounds and the arguments centred on the difference between the so-called organic and inorganic compounds. These terms had their origin in the classification, by the Swedish chemist Berzelius, of all the known chemical substances into the two groups, *organic* and *inorganic*. Substances like salt and water, coming from the inanimate world of air, soil and ocean, were classified as inorganic, while any compound derived directly or indirectly from a living organism was classified as organic. It was held that only living organisms could manufacture organic chemicals. Furthermore, although it was recognized that inorganic compounds obeyed a set of 'laws' governing their behaviour, it was believed that when the very same elements were present in organic compounds, they obeyed quite different laws. At that time most scientists agreed that the chemistry of living organisms was distinct from that of the inanimate world, and the concept of a 'vital force' was invoked to try to explain phenomena that were encountered only in living animals or plants.

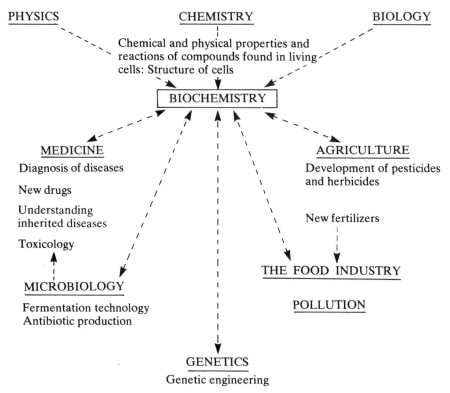

Figure 1.1 Relationships of biochemistry with other sciences

"I can make urea without needing a kidney"

In 1828, Fredrick Wöhler, one of Berzelius' students, started a revolution of ideas. He heated an *inorganic* compound, ammonium cyanate, and produced an *organic* compound, urea:

$$NH_4OCN \longrightarrow H_2N-\overset{\overset{\displaystyle O}{\|}}{C}-NH_2$$

ammonium   urea
cyanate

Urea was well-known and had been isolated from urine in 1773. However, since it was an *organic* compound, according to the vital force theory it could only be produced by living kidneys—hence Wöhler's statement, above.

At the time there was disagreement and controversy about Wöhler's discovery, but there could be no doubt that the result of the experiment was correct and soon other chemists were synthesizing organic compounds from inorganic ones. Eventually the vital force theory had to be abandoned as it gradually became accepted that properties possessed by living organisms could be explained in chemical terms. The same physical and chemical laws applied equally to living and non-living matter.

Today the term 'organic chemistry' simply refers to the branch of chemistry dealing with the compounds of the element carbon, whether these compounds are made by living organisms or by chemical reactions in the laboratory. Because of its atomic structure the element carbon can combine with other elements, especially hydrogen, oxygen, nitrogen and sulphur, and with itself, to form an enormous range of compounds. Over two million different organic compounds are known today, and the majority have been synthesized in the laboratory, not in living organisms.

What distinguishes living from non-living?

It is easy enough to observe the differences between living and non-living things. Living things breathe, feed, move, grow, reproduce, and so on. If there is no difference between organic compounds synthesized in the laboratory and the same organic compounds occurring in living organisms, and if there is no such thing as 'vital force', is there any *chemical* difference between the living and non-living? All living systems are made up of 'lifeless' molecules, so what leads them to have the properties we associate with the phenomenon we call life? The science of biochemistry developed because scientists asked questions such as these, and is based on the belief that it is indeed possible to give a *chemical* explanation for the above-mentioned biological properties, or *life activities*. This is how biochemistry came to be at the meeting place of biology and chemistry. Not only did scientists wonder: they also carried out experiments to try to find chemical explanations for these life activities.

What is biochemical research?

What experiments can be done to find out how living things work chemically? The first thing to be done is obviously to make chemical analyses of cells and tissues in order to find out what types of molecules are present. The methods of chemical analysis are therefore important fundamental tools of the biochemist. In the early part of the present century this was a major activity of biochemists, but it is no less important today, although many of the methods and techniques have changed dramatically. A key discovery was the realization that many of the 'molecules of life' were extremely large. They have been christened 'macromolecules' to signify this. We now know that life processes are vitally dependent upon macromolecules of various types, and an important part of modern biochemistry is the analysis of the structure of these giant molecules. One example is the compound responsible for the red colour of the blood, haemoglobin. This is a protein with a molecular mass of about

67 000 and the molecule contains something like 12 000 atoms. Its molecular formula is $C_{2952}H_{4664}O_{832}N_{812}S_8Fe_4$ and amazingly we know almost exactly where every atom in the molecule is placed. Haemoglobin is only of average size for a macromolecule: some macromolecules are very much larger. The whole of Chapter 2 is devoted to macromolecules and we shall see how highly-sensitive techniques have been developed to analyze the structures of these molecules.

The next question to ask is how these different organic chemicals, including the macromolecules, interact with one another to produce the phenomenon we call 'life'. Biochemists therefore study the reactions of the different types of molecule under the conditions found in living organisms. This is important as living organisms operate at comparatively low temperatures—not more than 40 °C for most of them—and within a quite narrow range between the acid and alkali regions of the pH scale. From the 1920s biochemical research was concerned with these reactions and interactions, and especially with the activities of the biological catalysts called *enzymes* which had also been found to be macromolecules. The reactions and interactions of the different molecules in cells came to be called *metabolism* which in turn is controlled by enzymes. The study of metabolism includes trying to understand how materials taken from the environment are converted into the materials found in cells, finding out how food reserves are built up and used at the appropriate times, and especially discovering how energy is obtained and used to drive all the various life processes. In this area of research the biochemist attempts to relate the chemical *structures* of the molecules of life to their biological *functions*. Chapters 3 and 4 are devoted to metabolism, and Chapter 3 is especially concerned with how living organisms obtain energy.

Much more recently biochemists began to study the mechanism by which our characteristics are passed on from generation to generation. This area is called *genetics*. The belief that it is possible to provide a chemical explanation for biological phenomena is nowhere better demonstrated than in this area. In this case, it is a matter of finding out how *information* is stored, how it is used and how it is passed on at cell division. The great revolution came in 1953 with the discovery by Watson and Crick of the structure of the complicated macromolecule deoxyribonucleic acid, now universally identified simply by three letters—'DNA'. This heralded the birth of 'molecular biology' and has given rise to such things as 'genetic engineering'. Chapter 5 is concerned with this aspect of biochemistry.

Where next?

Biochemistry has had a relatively short history in comparison with chemistry, physics, mathematics, medicine and astronomy. (It is interesting to look at the discoveries of one of the first 'biochemists', Louis Pasteur, although he would not have recognized the term (Box 1.1), and to see how they range from the purely chemical to the essentially medical.) Borrowing from these other scientific disciplines, biochemistry has progressed in leaps and bounds to become a fully-fledged science in its own right. It touches a great many areas of present day life—medicine, agriculture, the food industry and so on—and in the future its sphere of influence will surely continue to expand (see Chapter 6).

Box 1.1 The discoveries of Louis Pasteur (1822–1895)

1844	Separates isomers of tartaric acid by picking out crystals by hand	*CHEMICAL ANALYSIS*
1857	Demonstrates that lactic acid fermentation is caused by a living organism	*CHEMISTRY OF LIVING ORGANISMS*
1859	Shows that if airborne organisms are excluded, boiled broth does not putrify	*LIVING ORGANISMS RESPONSIBLE FOR PUTREFRACTION*
1881	Vaccinates sheep with living but weakened (attenuated) strain of anthrax bacilli	*USING KNOWLEDGE OF BIOCHEMISTRY AND MICRO-BIOLOGY*
1885	Vaccinates nine-year-old Joseph Meister with rabies vaccine. Joseph was still alive in 1940	*IN THE SERVICE OF MEDICINE*

For the moment we must go back (as it were to join the biochemists of the early part of the century) to ask what elements and compounds are found in living organisms.

1.2 The elements of life

The majority of the chemical elements present in living organisms are at the top of the periodic table, that is, mainly the lighter elements are found in living matter.

Those elements which are capable of existing in combination as organic compounds are plentiful in cells. (From this point of view it matters little what sort of cells we are talking about, for the elemental composition of nearly all cells is practically identical.) Such elements include carbon, oxygen, hydrogen, nitrogen, phosphorus and sulphur. Other elements such as sodium and potassium are also abundant but usually exist in organisms as *ions* rather than within compounds. These elements constitute the *bulk elements and ions* of living matter. Other elements, although equally important and vital for life, are found in much smaller quantities: these are the *trace* elements (Table 1.1). Not all of these are essential for all organisms and the list is in any case probably incomplete. Trace elements are usually found in combination with large organic molecules, often with proteins. One example is the element iron in haemoglobin. If you look back to the elemental composition of this protein (p. 41) you will be able to calculate that the percentage of iron is less than one-third of one per cent. However, this does not mean that it is unimportant. It is known that the function of the iron atoms is concerned with binding oxygen. For many trace elements, however, their exact function has yet to be discovered. In some instances we are merely aware that if they are not available to an organism then disease and death result.

Why have some elements been selected?

Does the abundance of elements in living organisms bear any relation to their

Table 1.1 The elements of life

Elements	Role
BULK ELEMENTS	
Carbon, Hydrogen, Nitrogen, Oxygen, Phosphorus, Sulphur	Universal in organic compounds of the cell
BULK IONS	
Sodium, Chloride, Potassium, Calcium	Water balance, osmotic pressure and acid–base balance, nerve conduction, bone
TRACE ELEMENTS	
Magnesium	Chlorophyll, enzyme cofactor
Iron	Haemoglobin, cytochromes
Copper	Enzyme cofactor
Molybdenum	Enzyme cofactor
Zinc	Enzyme cofactor
Cobalt	Constituent of vitamin B_{12}
Manganese	Enzyme cofactor
Boron, Silicon, Vanadium, Selenium and possibly others	Required for at least some organisms
Iodine	Constituent of thyroid hormone

abundance in the earth's crust? A look at Table 1.2 seems to show that it does not: certain elements have been selected and others rejected. The abundance of carbon in the earth's crust is about 0.2% and yet carbon may constitute up to 25% of the matter in living organisms. In contrast to this, silicon is extremely abundant in the earth's crust but is only present to a rather limited extent in living tissues. The reason for this is probably that only carbon can form the range of complex organic compounds upon which life is based. More generally, however, it may not be the abundance of an element so much as its *availability* that determines whether it is chosen to participate in life processes. (Of course, once it has been 'selected' at some distant time in the history of the earth, there is the strong likelihood of its continuing to be used.) Availability relates to whether or not an element can exist in water-soluble compounds or complexes in the earth's waters where life evolved.

1.3 The compounds of life

It was mentioned earlier, when the differences between living and non-living matter were being discussed, that the so-called organic compounds, originally isolated from plants and animals, were all compounds of the element carbon. Subsequently, organic chemistry was defined as the chemistry of carbon compounds. During the evolution of life, the versatility of carbon in forming such an enormous range of compounds was exploited to the full.

Table 1.2 Relative percentages of elements found in living organisms compared with the earth's crust

Element	Organisms	Earth's crust
Hydrogen	49.0	0.22
Carbon	25.0	0.19
Oxygen	25.0	47.0
Nitrogen	0.27	<0.1
Calcium	0.073	3.5
Potassium	0.046	2.5
Silicon	0.033	28.0
Magnesium	0.031	2.2
Phosphorus	0.030	<0.1
Sodium	0.015	2.5
Others	traces	13.7

The elements that are present in greatest abundance in all living cells are carbon, hydrogen, oxygen and nitrogen (Table 1.2). Much of the hydrogen and oxygen exist in combination as water. For the rest, these elements are found, along with smaller amounts of phosphorus and sulphur, in combination as organic molecules.

What sorts of organic compounds are found in living cells?

As a result of the research by chemists and biochemists in the early part of the present century, it became clear that all cells (plant, animal or microbial) contained four major classes of organic compounds: *proteins, carbohydrates, fats* and *nucleic acids* (Table 1.3). There were other compounds present, as well as water and inorganic ions, but these four classes of compounds were predominant and characteristic of living cells. Some of these molecules were very large, and at the time it was not possible to work out what their structures were. When they were broken down using gentle methods, however, it was found that they were all constructed from the same few basic building blocks. Thus all proteins, from whatever source, were built up from combinations of the same 20 or so amino acids, all nucleic acids contained the same 4 or 5 nucleotides, and so on. Their relationships are summarized in Table 1.3. Gradually, the structures of these building blocks and other small molecules from cells were discovered: some were component parts of macromolecules, others were compounds in the process of being incorporated into macromolecules, and yet others were food materials, vitamins, hormones and other participants in the process of metabolism. We shall be looking at them in greater detail in future chapters. The structures of the macromolecules themselves were not discovered until very much later. For example, it was not until 1953 that the complete structure of a rather small protein, insulin, was worked out and this was a major step forward.

All cells contain approximately the same range of organic compounds. Cells, and organisms, differ from one another because the 50–100 basic building blocks can be combined to make countless different types of macromolecules, each with its own characteristic properties and functions. Nevertheless, there is an underlying 'unity' in life: the processes going on in a bacterial cell are very similar to those going on in a human liver cell. This

Table 1.3 Substances found in cells and the building blocks of which these are composed

Substance	Percentage of total weight	Role	Constituents
Water	70		
MACROMOLECULES			
Proteins	15	Enzymes, structural, antibodies, etc.	20 types of amino acid
Polysaccharides	3	Storage of energy, structural	10–20 types of sugar
Nucleic acids	7	Storage and transmission of information	4 nitrogen-containing bases and sugar-phosphate backbone
SMALL MOLECULES			
Lipids	2	Storage of energy, insulation, structural	Fats, fatty acids
Metabolic intermediates	1–2	Stages in the break-down or build up of other molecules	Thousands of different compounds
Cofactors	<1	Participants in enzyme reactions	Vitamins, and many other substances
INORGANIC IONS See Table 1.1	1		

makes biochemistry much easier to understand than would otherwise be the case. For example, when a biochemist discovers a certain type of molecule in a given type of cell, he will, from his background knowledge of other cells, have a fairly good idea of what that molecule is doing, what it was made from and what its ultimate fate might be. Not only are the basic chemical structures of the compounds of life similar in widely-different species, but also the basic chemical processes by which these molecules interact with one another are very similar too.

Although life is based on carbon compounds, it is no less dependent on the 'solvent for life', namely water, which is the major constituent of all living cells.

1.4 Water: the solvent for life

All life goes on in a watery environment: life evolved in water and the major part of all living matter is water. It is true that some organisms can survive almost complete desiccation, but in this condition they exist in a state of 'suspended animation' with practically no life processes going on. Water is required not only as a medium in which the biochemical reactions of life can go on, but also for transporting substances in and out of cells, maintaining temperature, producing digestive fluids and secretions, and as a solvent for excreted waste products.

The special properties of water

The properties of the compound, water, are of crucial importance to living organisms. It is therefore important that biologists and biochemists understand these properties.

Water is not a linear molecule, but has its two hydrogen atoms set at an angle of 104.5° on the oxygen atom (Figure 1.2). Because of the electron-

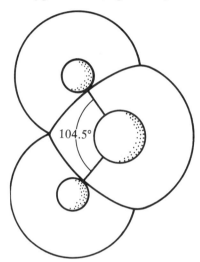

Figure 1.2 Structure of a molecule of water

attracting properties of oxygen, the electrons shared between the oxygen and hydrogen atoms tend to spend more time near the oxygen. This gives the molecule polarity with small net positive charges on the hydrogen atoms and a small net negative charge on the oxygen atom. Because of this uneven distribution of charge, nearby oxygen atoms of other water molecules are attracted to the hydrogen atoms to form weak bonds called *hydrogen bonds.* As the angle of 105° between the covalently-bonded hydrogen atoms is very close to the angle in a perfect tetrahedron (109° 28'), an almost tetrahedral arrangement can build up (Figure 1.3). This can involve an enormous number of water molecules, and although each individual hydrogen bond is rather weak, the net result is that a body of water has quite different properties from other hydrides such as hydrogen sulphide. This unique property of hydrogen bond formation results in water existing as a liquid at the normal temperatures of the surface of the earth.

Because of the non-uniform charge distribution on the water molecule, water has a high dielectric constant. It therefore acts as an electrolytic solvent. This means that ions of dissolved electrolytes such as sodium chloride can separate sufficiently to move independently as Na^+ and Cl^-. This confers on individual ions almost the properties of independent molecules because of the great reduction in electric field brought about between pairs of oppositely-

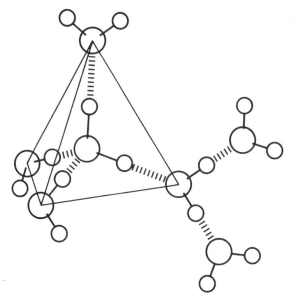

Figure 1.3 Hydrogen bonding between water molecules

charged ions (Figure 1.4). The same will be true of large charged organic molecules. Water itself dissociates to a small extent and this has important consequences in biological systems (Box 1.2).

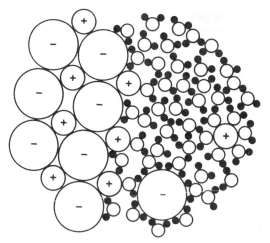

Figure 1.4 Water acts as a solvent

In a body of water, the formation of lattices of tetrahedral structures gives an 'ice-like' structure. This ice-like structure persists even in liquid water and results in properties of great biological importance. In the lattice structure the hydrogen bonds are constantly breaking and reforming, but even at 25 °C a considerable proportion of a body of water, on average, possesses this structure. Water has a high specific heat capacity and this is vital in the thermal

Box 1.2 Ionization of water and the pH scale

Water, H_2O, dissociates to a very small extent:

$$H_2O \rightleftharpoons H^+ + OH^-$$

The hydrogen ion (H^+) concentration in pure water is 0.000 000 1 mol l^{-1}. For convenience, this is usually written as the negative logarithm of the concentration and called the pH value.

$$H^+ \text{ concentration} = 0.000\ 000\ 1 \text{ mol } l^{-1}$$
$$\log 0.000\ 000\ 1 = -7.0$$
$$-\log 0.000\ 000\ 1 = 7.0 = \text{'pH'}$$

The pH scale runs from 0 to 14. Solutions with an abundance of H^+ have pH values between 0 and 7 and are classified as *acids*. Acids are compounds that can dissociate to yield H^+. Thus, hydrochloric acid, HCl, gives H^+ and Cl^-. The stronger the acid, the higher the H^+ concentration and the *lower* the pH.

Solutions with pH values over 7 are said to be *basic* or *alkaline*. A base is a substance that dissociates to yield hydroxyl ions (OH^-) or that can accept H^+. Thus sodium hydroxide dissociates to Na^+ and OH^-. Bases reduce the H^+ concentration, but *raise* the pH.

←——*HYDROGEN ION CONCENTRATION INCREASING*

pH   0 1 2 3 4 5 6 7 8 9 10 11 12 13 14

MORE ACIDIC   NEUTRAL   MORE BASIC

The pH scale may be drawn as in the diagram above. However, remember that the scale is logarithmic and that a change of 1.0 of a pH 'unit' means a ten-fold change in the H^+ concentration.

The pH of some fluids

	pH
Blood	7.4
Gastric juice	1.2–3.0
Cow's milk	6.6
Urine	5.0–8.0
Grapefruit juice	3.2

stabilization of living organisms and in the stabilization of large masses of water such as seas and lakes. Another key property of water is the possession of a high latent heat of evaporation. Even near boiling point many hydrogen bonds exist which have to be broken before the vapour phase of water can form. This high latent heat of evaporation is important in humans in the process of sweating, which helps to maintain a steady body temperature.

We shall meet other important properties of water later in this chapter, but first we return to the question of how any particular cell obtains its supply of raw materials and energy from the environment.

1.5 Nutrition: living off the environment

One of the life activities mentioned previously was *feeding*. This means different things to different organisms. Any cell has to obtain a supply of raw materials and of energy. Raw materials are needed to build up new cell substance, and energy is required for this building up process as well as for other activities such as movement and osmotic work. Raw materials and energy come from the cell's environment, but it is clear that some organisms have very complex nutritional requirements while others are capable of living on very simple compounds. Why is this?

Why are the nutritional requirements of some organisms complex and some simple?

During the course of evolution different ways of obtaining raw materials and energy developed. The very first organism obviously had to make do with simple substances. Many bacteria and all green plants today still require only carbon dioxide, water, ammonia and certain mineral elements, and from these they build up all the complex molecules found in cells. The energy to do this comes either from sunlight, or in the case of some of the bacteria, from oxidizing certain compounds such as sulphite to sulphate. This mode of nutrition is called *autotrophic*.

Very soon in the course of the history of the earth, however, other organisms evolved whose mode of nutrition was to eat other organisms. The ability to use the very simple compounds was lost, and instead there was a requirement for a complex mixture of organic compounds. This mode of nutrition is called *heterotrophic*. Energy now comes from oxidizing the organic substances obtained by eating other organisms: the raw materials or building blocks for growth also come from the same source. Heterotrophic organisms are not necessarily highly complicated ones. Humans and all animals are heterotrophic because they lack the ability to obtain energy from sunlight. Equally, however, many bacterial cells are heterotrophic, for example, those which infect animals and plants, living as parasites or saprophytes. If we put the various pieces of the nutritional jigsaw together we discover that the bulk elements of life are being recycled and that the only major input of energy into the system is light from the sun. Another way of expressing this is to say that there is a food chain or food web. Organisms at the bottom of a food chain are autotrophic: all the rest depend on them. The bacteria and fungi that cause what we call 'decay' are equally important. By breaking down dead animal and plant tissues they complete the recycling process as well as satisfying their own nutritional requirements at the same time. Nevertheless, but for them we would be up to our necks in dead material!

1.6 The biogeochemical cycles

Heterotrophic organisms live at the top of a pyramid being supported by a whole range of autotrophic organisms. Of these autotrophs, some are *photo-autotrophs* such as the green plants and photosynthetic bacteria that use light energy. Others are *chemoautotrophs* which obtain energy not from light but from oxidizing simple compounds they find in their environment—such as sulphite to sulphate.

When an organism dies its tissues start to be degraded by the action of micro-organisms and the simple compounds produced as a result of this

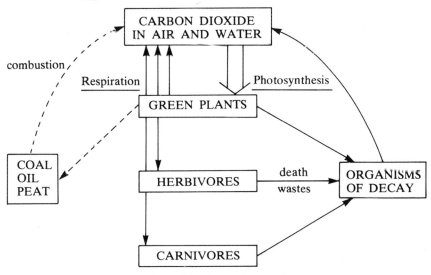

Figure 1.5 The carbon cycle

process of decay are made available for yet other organisms to use. In this way the elements of life are recycled.

The carbon cycle

A simplified form of the carbon cycle is shown in Figure 1.5. The amounts of carbon taking part in these processes are enormous as is shown in Table 1.4. Initially, carbon from carbon dioxide is incorporated into organic compounds during photosynthesis. This carbon is transferred to herbivores when they eat plants and then to carnivores when they eat herbivores, and so on. During these transfers, a single carbon atom may move through a variety of organic compounds in the course of metabolism. Eventually, energy to sustain and propagate life will be extracted from the organic compounds which in turn become oxidized to carbon dioxide in the process called respiration.

Table 1.4 Amounts of carbon taking part in the carbon cycle

Photosynthesis by green plants and bacteria results in:
      the annual synthesis of 50 000 000 000 tonnes of organic carbon compounds
      the annual release of 130 000 000 000 tonnes of oxygen
In the process 200 000 000 000 tonnes of carbon dioxide are taken from the atmosphere

Combustion of fossil fuels by man liberates:
      3 000 000 000 tonnes of carbon dioxide annually—
      less than 1% of the above amount of carbon dioxide used by organisms

A small amount of the carbon participating in the carbon cycle may for a time become immobilized in deposits of coal or oil. Such reserves of carbon are formed when plant and animal material becomes buried before complete decomposition can take place. Some three hundred million years later man may come along and dig up the coal or drill for the oil and burn these, thus releasing carbon dioxide into the atmosphere to take part in the carbon cycle once again.

The oxygen cycle

The oxygen of the atmosphere is constantly being renewed by photosynthesis, but it is believed that the earth's atmosphere originally contained no oxygen. Only when photosynthetic organisms evolved some two thousand million years ago did free oxygen appear in the atmosphere.

In the upper layers of the earth's atmosphere there is a 10 mile thick layer containing ozone. This substance does not participate in life processes, but has an important function nonetheless. Ozone has the property of absorbing ultraviolet radiation which in excessive amounts would be harmful to living things on the earth's surface.

The energy cycle

In the relationship between carbon and oxygen there is a third participant, namely energy. Photosynthesis involves trapping light energy and respiration involves releasing energy by oxidation of organic carbon compounds. Energy arrives at the surface of the earth as light. It may be trapped and recycled a number of times and eventually will be converted to heat. As carbon flows through the food chains only about 10% of the energy obtained is stored in an organism's tissues and is therefore available for transfer to the next organism. The remaining 90% is transferred to the surroundings as heat. In view of this loss of energy as heat at each step in a food chain, it is clear that any shortening of the food chain will minimize energy losses. It is a problem in the world today that humans in the developed countries are carnivores (really omnivores) eating animals fed on plant material. In contrast, those living in the developing countries are to a large extent vegetarians eating plant material directly. When individuals in affluent countries feed grain to fatten beef cattle they are using, in an inefficient way, food resources that could feed the hungry of the third world.

The nitrogen cycle

There is an abundance of nitrogen in the atmosphere: about 78% of air is nitrogen. All living organisms require nitrogen because it is a component of amino acids, proteins, nucleic acids and many other compounds of life. However, the nitrogen of the atmosphere is unreactive and is not capable of being used directly by the majority of life forms. It is first necessary to turn the atmospheric nitrogen into usable compounds and this process is called *nitrogen fixation*. Some nitrogen fixation occurs as a result of electrical discharges such as lightning (Figure 1.6), and even more occurs in the modern world as a result of industrial fixation during the commercial production of fertilizers. However, of the one hundred million tonnes of nitrogen added to the surface of the earth every year only about 10% is in the form of artificial fertilizer. The rest is added as a result of nitrogen fixation by certain microorganisms including a few types of bacteria and many of the blue-green algae

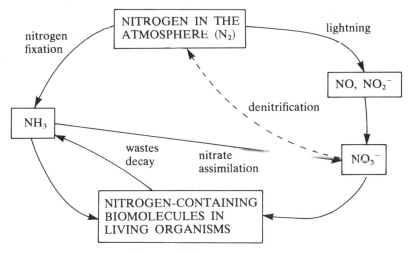

Figure 1.6 The nitrogen cycle

(see Chapter 4). Some of these live free in soil or water while others live in symbiotic relationships with green plants in nodules on the roots (Figure 1.7). Bacteria living in nodules not only fix sufficient nitrogen for their own needs, but also produce an excess of fixed nitrogen which they pass on to the host plant, largely in the form of amino acids. In addition, a surplus may frequently be produced and excreted into the surrounding soil. Farmers often deliberately build up the nitrogen content of their fields by periodically planting legumes. One hectare of alfalfa, for example, may fix up to 500 kg of nitrogen.

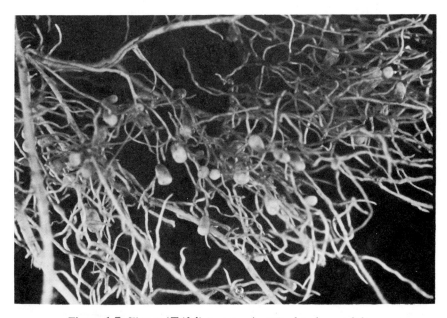

Figure 1.7 Clover (*Trifolium repens*) roots showing nodules

Phosphorus and other minerals

Phosphorus is an important element, especially for the formation of nucleic acids, and this and other elements are cycled to some extent. Along with nitrogen, phosphorus is one of the chief ingredients of commercial fertilizers. However, unlike nitrogen and carbon the natural reservoir is the rocks and not the atmosphere (Figure 1.8). Phosphate from rock leaches out very slowly and the dissolved phosphate becomes available to plants which pass it to animals.

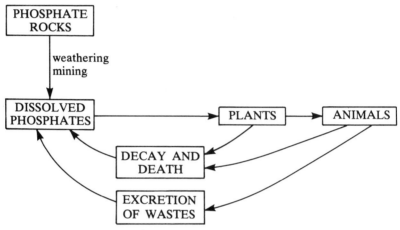

Figure 1.8 The phosphorus cycle

Biogeochemical cycles can be upset, often as a result of man's activities. The consequences may be serious ecological problems, destruction of plant life, and health problems for animals and man (Box 1.3).

Box 1.3 What happens when the biogeochemical cycles are disturbed?

Because of their cyclic nature, a disturbance or lack of balance at one point in a cycle may seriously affect the operation of the whole cycle. There are (unfortunately) only too many examples in the world today. It is possible for such disturbances to occur by natural causes: more often they happen as a result of human activity.

An example is phosphorus availability. Because of the mining of several million tonnes of phosphate rock each year, the rate of movement of phosphate from rocks to the waters of the earth has accelerated. Normally, availability of phosphate is the limiting factor for algal growth. With the increase from phosphate in sewage, detergents and run-off from fertilizers, many freshwater lakes have experienced an *algal bloom*, with a dramatic increase in growth rate. At the end of the growing season the algae die and sink to the bottom, stimulating a massive growth of bacteria the following year. The oxygen at the deeper levels becomes depleted with the result that fish cannot exist. Chemical changes resulting from the deoxygenation of the water can produce odorous, toxic gases from the bottom sediment, further decreasing its desirability as a habitat.

Such 'ecological disasters' are, of course, avoidable, but at what cost? Improved sewage treatment is one way of dealing with the problem, but it involves millions of pounds in the development of tertiary treatment of sewage. It would, of course, help if the phosphate content of detergents could be reduced, but much of the phosphate now comes from run-off from fertilizers. The use of these is essential if agriculture is to produce enough food for a growing population. There is no cheap or easy solution.

1.7 Membranes and cells

We have seen that life depends upon the reactions and interactions of complex organic chemicals, including macromolecules, in a watery environment. However, if these molecules are simply dispersed in a random fashion in an aqueous medium they do not themselves show the properties we associate with 'life'. The non-living molecules must be *organized* in the right way for life activities to occur.

The cell theory

One of the most important discoveries to result from the invention of the optical microscope in the seventeenth century was that the tissues of living things were made up of cells. This was an important landmark in biology and was the start of the so-called *cell theory*. According to this theory, all living organisms are composed of cells, and, in addition, all cells arise from other cells. The development of this theory signalled the recognition by scientists that living material was highly organized and that the cell was the basic unit of structure and function. It was not until much later that the development of the electron microscope allowed the interior organization of cells to be observed, revealing an even greater degree of complexity, but which could nevertheless be related to biological function. We can now define living organisms as *chemical organizations composed of cells and which are capable of reproducing themselves.* (We shall see later that it is sometimes difficult to draw a sharp line between the living and the non-living when we come to look at viruses—but the above statement is a reasonable 'working definition'.) If the aim of the biochemist is to understand the chemistry of living organisms, he must also understand why it is necessary to have this high degree of organization in cells. Before we consider the intricate internal structure of cells, we pause here to consider why the cell is the fundamental unit of structure of organisms.

Why did cells evolve?

How did the cell come to evolve? Life arose on earth thousands of millions of years ago when collections of organic molecules in the 'primaeval soup', which formed the primitive ocean, became isolated from the bulk water of that ocean. Because of their closeness, these molecules were able to interact with one another and began to show the first signs of life. But the close proximity of the molecules could only be maintained if there was some sort of barrier enclosing the molecules and separating them from the environment. In the absence of such a barrier, the molecules forming the 'living material' would come to equilibrium with the surroundings. In practice this means that any molecules concentrated into one locality would tend to drift apart, eventually becoming too far apart to interact with one another. Life is not at equilibrium with the surroundings, or, to put it another way, living structures only come to be at equilibrium with the environment when they are dead! In order to maintain this 'non-equilibrium' it is necessary to have a barrier between the living matter and the environment. If we ask the question how barriers can be built we should by now expect the answer to be a *chemical* one because biochemists seek chemical answers to biological problems.

Hydrophilic and hydrophobic interactions

Some compounds, like salt, are soluble in water while others, like oil, are not. The latter try to keep away from water by forming globules with the smallest possible surface area exposed to the water. Many compounds are soluble in water because they can form *hydrogen bonds* with the water molecules. Oily compounds, by contrast, do not form hydrogen bonds with water molecules and are insoluble. Compounds that can form hydrogen bonds with water are called *hydrophilic* or water-loving compounds: compounds that cannot are called *hydrophobic* or water-hating ones. When water molecules surround hydrophobic compounds they can maintain their ice-like structure, and to break this up requires the expenditure of energy. This is why the hydrophobic molecules tend to come close to one another and keep separate from the water.

Now let us imagine what would happen if we constructed a molecule hydrophilic at one end and hydrophobic at the other (Figure 1.9). Such molecules are widespread in both biological and chemical systems. When dispersed in water, molecules of this type can form droplets or 'micelles' with the hydrophobic tails pointing 'inwards' and the hydrophilic heads pointing 'outwards' to the watery phase. Another possible arrangement is more interesting, however, because using such molecules it is possible to form a *double layer* (Figure 1.9). In this arrangement the hydrophobic tails point to one another and the hydrophilic heads point into the watery phase. Now we can use such an arrangement to keep two watery phases apart, and we can construct a barrier with complex organic molecules dissolved in water 'inside' (cell contents) and watery phase 'outside' (environment). In this way we have constructed a primitive 'cell' and the double layer of hydrophilic–hydrophobic molecules surrounding it is a primitive *membrane*. The biological molecules involved in the formation of membranes are the lipids, especially the phospholipids. As we shall see, all cells are surrounded by a membrane, sometimes called a *plasma membrane*, and many cells have, in addition, a *cell wall* which is a supporting rather than a separating structure.

Communication through the barrier

For life activities to be possible it is absolutely essential to surround our collection of organic molecules by a membrane. It is important that such a barrier is not completely impermeable. If the living organism is to survive, it must take in raw materials and energy from the environment and excrete waste products into it. There must be uptake of oxygen for respiration and elimination of carbon dioxide. Therefore the basic membrane structure must be modified so things can pass in and out *selectively*. It seems that this modification is brought about by embedding in the membrane structure,

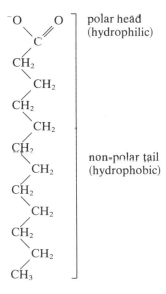

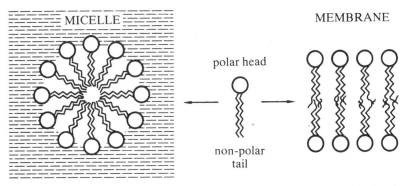

Figure 1.9 A long-chain fatty acid molecule is hydrophilic at one end and hydrophobic at the other. In an aqueous environment such molecules tend to form micelles or bilayers

protein molecules which in various ways aid the transport of molecules and ions into and out of cells. The proteins can exist either on the 'inner' or on the 'outer' surface of the membrane, or may span the whole membrane forming selective channels. Such an àrrangement of proteins in a membrane is shown in Figure 1.10: it is in many ways a hypothetical picture, but explains many observations on the properties of membranes. The protein molecules can be imagined as being something like icebergs floating about in the sea of lipid molecules. This idea is important because it is believed that the cell membrane is a constantly-changing dynamic structure rather than a rigid, static structure. Certainly membranes are highly selective in what they allow in and out of cells. A cell in sea water, for example, can expel sodium ions, although these are in high concentration in sea water, and allow in potassium ions, which are in low concentration.

Why are cells the basic unit of biological structure?

Every cell is a self-contained and, at least in part, a self-sufficient unit surrounded by a membrane that controls the passage of materials in and out. This makes it possible for the cell to differ chemically from its surroundings. Before considering the structure of cells, we must ask why cell size is restricted. Although there is a great variety of sizes of cells, most plant and animal cells are between 10 and 30 µm in diameter. The reason for this is that materials entering and leaving cells must pass through the cell membrane, which is a surface. The more active the cell, the more rapidly oxygen and food materials must pass in and carbon dioxide and waste materials pass out. The smaller the cell, the greater the ratio of surface area to volume: for a sphere, the surface area increases as the square of the radius while the volume increases as the cube. It is no surprise therefore to find that cells with a high rate of metabolism such as bacteria are very small. It may be argued that some cells are very large—for example, an ostrich egg yolk is a single cell and may be several centimetres across. Such cells are not highly active, however. A high rate of metabolism commences after fertilization, but then practically the first thing to happen is a series of cell divisions which produce a number of *small* cells. The effect of this is to increase the area of membrane for a given mass of tissue, thus making for more rapid transport in and out.

1.8 The structure of cells

Practically all organisms are built up of cells. The majority of these cells are too small to see with the naked eye and the development of the 'cell theory' had to await improvements in the construction of optical microscopes. In the same way, the discovery of the internal structure of cells had to await the development of the electron microscope. Table 1.5 gives the dimensions of cells in comparison with their organelles ('little organs') and with some of the molecules which they manipulate.

Table 1.5 Dimensions of cells, organelles and molecules

CELLS	
Radius of ostrich egg	20 cm
Radius of giant algal cell	5 mm
Cross-section of squid giant nerve cell	1 mm
Radius of *Amoeba*	0.1 mm
Most cells in range { eukaryotes	20 µm
Most cells in range { prokaryotes	1 µm
ORGANELLES	
Chloroplast	8 000 nm
Liver mitochondrion	1 500 nm
Ribosome	20 nm
MOLECULES	
Haemoglobin	6.8 nm
Glucose	0.7 nm

As information about cells from different types of organisms accumulated it became possible to classify all organisms according to their basic cell types as *prokaryotic* or *eukaryotic*, to be described below. The basis of this classification is very simple although it turns out that there are, in fact, many differences between the two groups. Those organisms whose cells do not have a well-defined nucleus are called prokaryotes (Greek: *pro-* before, *karyon-* nucleus). The organisms in this group are almost all unicellular and comprise the bacteria (Box 1.4) and the blue-green algae. All other organisms have cells which possess a definite nucleus and are called eukaryotes (Greek: *eu-* true).

Box 1.4 The bacteria

Bacteria form the oldest and most abundant group of living organisms. They are found in practically all environments from near-boiling hot springs to dark, cold ocean depths. Some varieties are actually killed by oxygen (obligate anaerobes), others can take it or leave it (facultative anaerobes), and yet others live only aerobically. Some are photosynthetic.

A single gram of soil may contain 2×10^9 bacterial cells, and soil bacteria are especially important ecologically as fixers of atmospheric nitrogen (Chapter 4). A few varieties of bacteria cause disease in man and other animals, but others live in the digestive tract where they may actually be essential to the survival of the host animal. Some bacteria are useful and indeed important commercially, socially and medically, e.g. as agents in the production of cheese and vinegar, as participants in sewage processing and as producers of antibiotics.

(a)   *(b)*   *(c)*

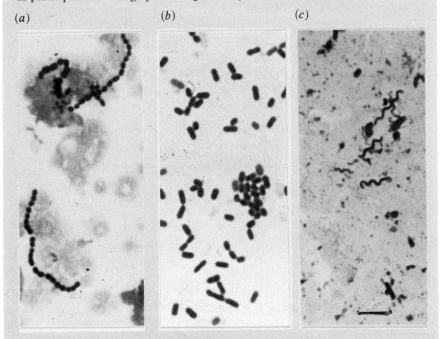

Some typical bacteria as seen under the microscope: (*a*) *Streptococcus pyogenes*, (*b*) *Escherichia coli*, and (*c*) *Spirillum* species. Length of line 5 μm. (*Photographs kindly supplied by A. P. D. Wilcock, Department of Microbiology, University of Leeds.*)

The biological success of the bacteria is due to their extreme versatility and adaptability coupled with their rapid rate of cell division. They have served biochemists very well as experimental organisms, especially in the areas of metabolic study and molecular biology, as we shall see. The major groups of bacteria are:

Eubacteria: Rod-shaped, spherical, curved or spiral forms; some mobile with flagella; all possible modes of nutrition (chemoautotrophy, photosynthetic, autotrophy, heterotrophy); distributed in soil and water as decomposers and symbionts, but also pathogenic causing such diseases as tetanus, diphtheria and tuberculosis.

Myxobacteria: Rod-shaped, flexible, often gliding in slime; usually heterotrophic, soil-dwelling decomposers.

Spirochaetes: Very long, helical, twisting; heterotrophic; live in polluted water and as parasites or pathogens; cause diseases such as syphilis and relapsing fever.

Rickettsiae: Very small; non-mobile heterotrophs; often intracellular parasites; cause typhus.

Mycoplasma: Smallest free-living cells, but no cell walls; heterotrophic; live in soil but many parasitic and pathogenic; cause pneumonia.

As we have seen, all cells have a cell membrane, but cell membranes are not visible in the optical microscope. In the electron microscope, however, the membrane can be seen as a continuous double line around the cell, about 9 nm in thickness. According to the current theory, the membrane is a double layer of lipid with protein molecules embedded at intervals (Figure 1.10). Although there are differences in the types of lipids and proteins, no doubt related to differences in function, this basic structure is common to all living systems.

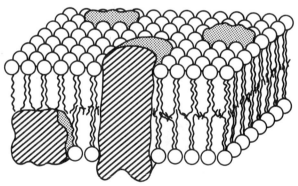

Figure 1.10 Proposed structure of a membrane—a lipid bilayer with molecules of protein embedded in it

Some cells, especially those from plants and bacteria, are surrounded by a cell wall. This is a rigid structure, distinct from the cell membrane. Its function is structural and it plays little part in controlling the entrance and exit of materials from cells. In plant cells, the cell wall is typically made of the polysaccharide cellulose, and is a supporting structure. In bacteria, the cell wall usually has a more complicated chemical structure (see p. 80) and acts more as a restraining net than a skeletal element.

Prokaryotes

Prokaryotic cells are typically very small, ranging in size from 0.1 μm to several μm in diameter. They frequently have appendages such as flagella for locomotion and pili which are shorter projections. The cells are often coated with a gelatinous capsule, and inside this is the rigid cell wall mentioned above. These outer structures are secreted by the cell, but under certain conditions the cell can live perfectly well without them.

Prokaryotic cells have very few inclusions or organelles compared with eukaryotes (see below). There may be granules of various sorts which appear and disappear depending upon the metabolic condition of the cell, such as granules of β-hydroxybutyric acid which some bacterial cells accumulate when carbon and energy sources are plentiful. Many bacteria form tiny resistant vegetative bodies called spores. Photosynthetic prokaryotes (the blue-green algae and the photosynthetic bacteria) have complex arrangements of membranes within the cell that contain the photosynthetic apparatus, but these have a different structure from eukaryotic chloroplasts.

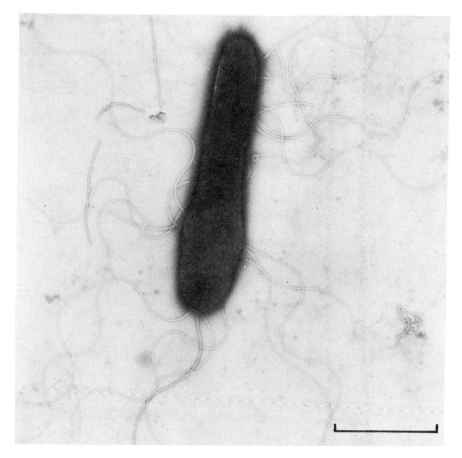

Figure 1.11 Electron micrograph of a bacterial cell (an unidentified cellulose-degrader from sewage) with peritrichous flagella. Length of line 1 μm. (*Photograph kindly supplied by Karen Walters*)

Apart from the photosynthetic apparatus and the nuclear body (see later) the only true organelles of prokaryotes are the ribosomes. These are small (20 nm), approximately spherical particles made of nucleic acid and protein, and are essential for the synthesis of proteins. Some ribosomes are free in the cytoplasm while others are attached to the cell membrane. The genetic apparatus of prokaryotes, usually a single piece of DNA, is to some extent folded, but it is not contained within a membrane as is the case in eukaryotes. The area of the cell containing the DNA is sometimes referred to as the 'nuclear body'. Figure 1.11 is an electron micrograph of a bacterial cell; Figure 1.12 is a diagram of a generalized bacterium. The viruses (Box 1.5) represent a distinct group of organisms.

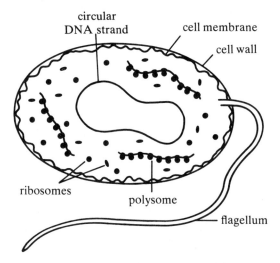

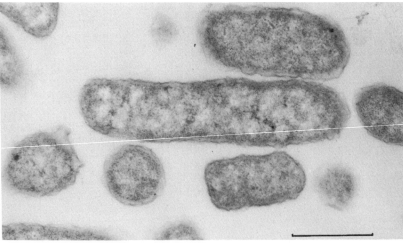

Figure 1.12 Diagram showing the structure of a typical bacterial cell. The electron micrograph shows a *section* through cells of a bacterium, *Escherichia coli*, commonly found in the intestine. Very few organelles are visible. Several cells have been sectioned at different angles by chance. Length of line 1 μm

Box 1.5 *The viruses*

On the borderline between the living and the non-living are the viruses. These are of great medical and social importance because of the diseases they cause to humans, animals and plants, from measles to foot and mouth disease.

Viruses are extremely small particles, typically composed of nothing more than a small piece of nucleic acid (DNA or RNA) surrounded by a protein coat called a *capsid*. A mature virus particle is called a *virion*. The size of viruses is such that they can only be seen in the electron microscope.

(a)   (b)

(c)   (d)

(e)   (f)

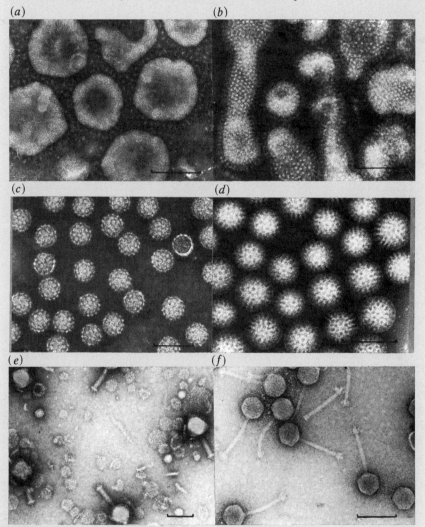

Electron micrographs of some viruses. (*a*) Coronavirus—one of the viruses that causes the common cold; (*b*) influenza virus; (*c*) wart virus, (*d*) rotavirus—a virus that causes diarrhoea in babies and infants, (*e*) and (*f*) bacteriophages— viruses that attack bacteria. Length of line 0.1 μm. (*Pictures a–d kindly donated by Dr. J. D. Almeida, The Wellcome Research Laboratories, and e and f by Dr J. H. Parish, University of Leeds.*)

By themselves viruses do nothing, that is, they perform none of the life activities we associate with living: moving, feeding, breathing, etc. Viruses only 'come to life' when they invade another cell, animal, plant or even bacteria. Once inside a living cell they take over and organize the metabolism of that cell so that it produces more virus particles. Eventually, the host cell may burst releasing several hundred virus particles—and the cycle begins again.

The piece of nucleic acid in a virus particle represents a few 'genes', that is, a few instructions which come into action when the virus has invaded a cell. The protein coat is probably protective, but may also aid the virus in getting its nucleic acid into the host cell.

Within a very limited size and number of constituent molecules, viruses have evolved a great variety of forms, from the rod-shaped tobacco mosaic virus, to the more or less spherical polio virus, to the highly complex 'T-phages' that infect bacterial cells. Sizes range from 25 nm diameter spheres (actually icosahedra) to 18×300 nm rods. The very smallest viruses contain only about 9 genes, but this is apparently sufficient to take control of the whole metabolic machinery of a cell.

Almost all types of cell and organism are attacked by viruses, yet individual varieties of virus are highly specific about the type of cell they invade. Many animal viruses are unable to attack human cells, and the bacterial viruses, or 'bacteriophages', will often only attack a single type of bacterial cell.

Size comparison

	Diameter (μm)	Volume (μm³)
Virus	0.025	0.006
Bacterial cell	1.0	1.0
Liver cell	20.0	4 000.0

Eukaryotes

An enormous number of types of eukaryotic cell is known and part of the reason for this is that many eukaryotes are multicellular organisms. Division of labour results in certain cells becoming highly specialized for certain functions. We shall concern ourselves here with the common features and with some typical organelles.

Flagella and cilia may be present, but their detailed structures are very different from those found in prokaryotes. For example, they are much larger (cilia $2–10 \times 0.5$ μm; flagella $100–200 \times 0.5$ μm) and are constructed from sets of comparatively rigid elements surrounded by an extension of the plasma membrane.

Eukaryotic cells typically contain many organelles, and we must remember that many of these organelles are larger in size than the average bacterial cell. We know a great deal about eukaryotic organelles, not only because their structures can be seen in the electron microscope, but also because these organelles can be isolated from cells by special techniques. The chief method that has been used is that of cell fractionation in which broken-up cells are spun in a centrifuge at different speeds. Because of their different sizes and densities the various organelles sediment towards the bottom of the centrifuge tube at different rates. By means of this technique the various organelles have been separated from one another (Figure 1.13), and this has made it possible for biochemists to study their biological function in isolation. Thus the reactions of photosynthesis may be studied with a preparation of chloroplasts.

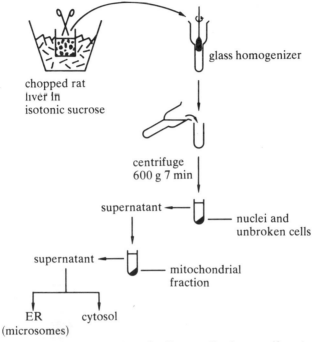

Figure 1.13 Fractionation of cell organelles by centrifugation

The following are the major organelles of eukaryotic cells, but a complete list would be very much longer. Plant and animal cells are compared in Figure 1.14.

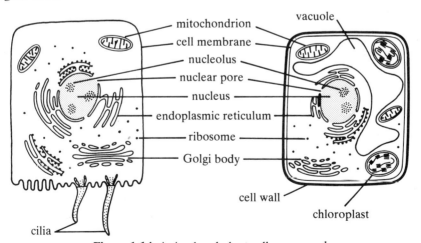

Figure 1.14 Animal and plant cells compared

Nucleus

The nucleus is the most prominent body in the cell and is surrounded by a double membrane (Figure 1.15). These membranes appear to form circular nuclear pores in certain places of 80–100 nm diameter. These may occupy up to one-third of the total surface area of the nucleus and are thought to be responsible for the selective passage of materials into and out of the nucleus. The chief component of the nucleus is DNA but RNA and protein are also present. Another structure often clearly defined within the nucleus is the nucleolus which contains a large number of granules rich in RNA. These granules are the precursors of the ribosomes. The nucleolus has no membrane of its own.

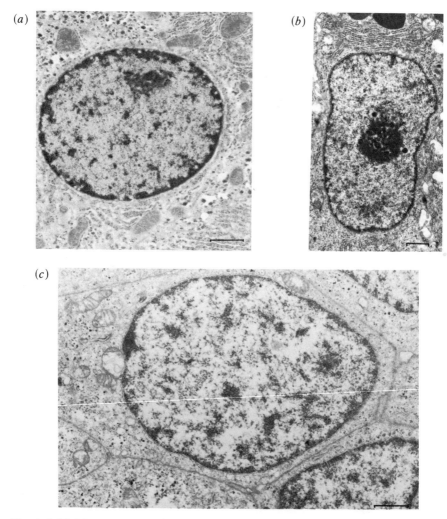

Figure 1.15 Electron micrographs showing the nuclear region of cells. Length of line in each case 1 μm. (a) Part of a liver cell; (b) part of a rat yolk-sac cell, and (c) a cell of human foetal kidney tissue where the nucleus is very large

Endoplasmic reticulum

In the cytoplasm of most eukaryotic cells there is a network of membranes called the endoplasmic reticulum. When the cell is disrupted fragments of this may be separated (see Figure 1.16) and are referred to as the 'microsomal fraction'. The cavities within the endoplasmic reticulum function as channels and reservoirs in internal transport and storage. Ribosomes are often found attached to the endoplasmic reticulum giving it a granular appearance. Such regions of endoplasmic reticulum are called 'rough endoplasmic reticulum' to distinguish them from 'smooth endoplasmic reticulum' which has no attached ribosomes. As in the case of prokaryotic ribosomes, eukaryotic ribosomes are responsible for the synthesis of protein. It is believed that proteins destined for export from the cell are synthesized by the ribosomes attached to the endoplasmic reticulum.

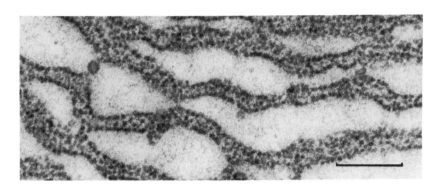

Figure 1.16 Endoplasmic reticulum. The electron micrograph shows the typical appearance of rough endoplasmic reticulum with ribosomes along the membranes. When the cell is broken up fragments of endoplasmic reticulum form small sacs or vesicles called 'microsomes'. Length of line 1 μm

Golgi apparatus

Another system of complex vesicles and membranes found in most eukaryotic cells is referred to as the 'Golgi apparatus' (Figure 1.17). Material produced within the cell for export is 'processed' within the Golgi apparatus and is packaged in pinched-off vesicles derived from it. Eventually the vesicles fuse with the plasma membrane and their contents are released to the exterior. The digestive enzymes of the pancreas are produced and released in this way.

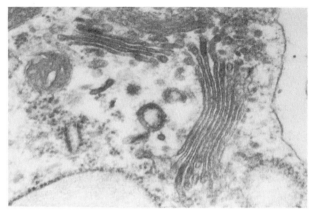

Figure 1.17 Golgi apparatus. Electron micrograph of a section of a cell showing the flattened sacs of the Golgi apparatus

Mitochondria

The oxidative processes of cell metabolism which are concerned with energy production from foodstuffs take place in the mitochondria. These may be spherical or elongated or even ramifying filamentous structures, but are always enclosed by a double membrane. The inner membrane is infolded to give characteristic structures called 'cristae' (Figures 1.18 and 1.19).

Mitochondria behave almost as if they were autonomous organisms within the cell. They have ribosomes, a protein-synthesizing apparatus, their own DNA and 'reproduce' by binary fission.

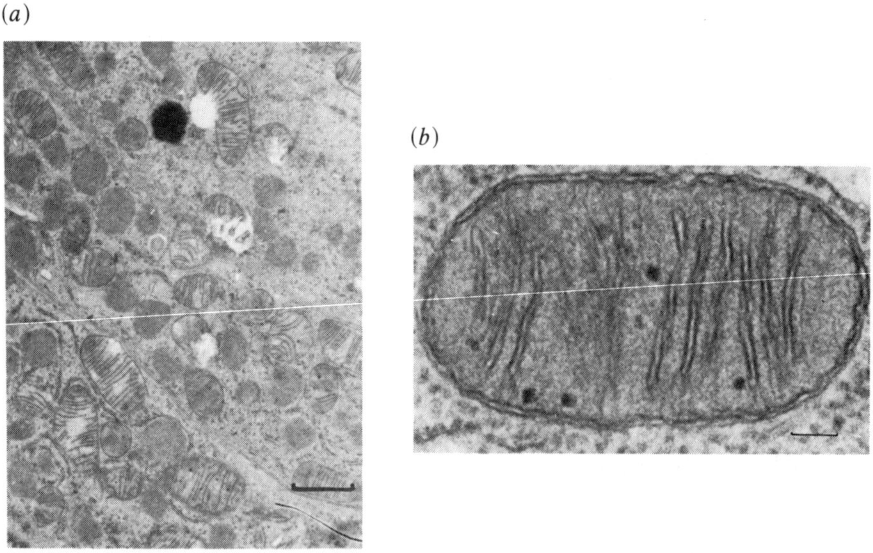

(a)

(b)

Figure 1.18 Electron micrograph of a section of a cell from rat pancreas showing (a) mitochondria, and (b) a single mitochondrion. Length of line (a) 1 μm, (b) 0.1 μm

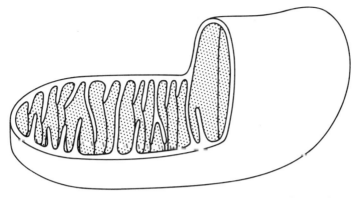

Figure 1.19 Cutaway diagram to show structure of mitochondrion and arrangement of cristae

Chloroplasts

Only photosynthetic cells possess the chlorophyll-containing structures called chloroplasts. These have a highly-characteristic structure of membranous vesicles contained within a plasma membrane. Like mitochondria they are semi-autonomous and have their own DNA and ribosomes (Figures 1.20 and 1.21).

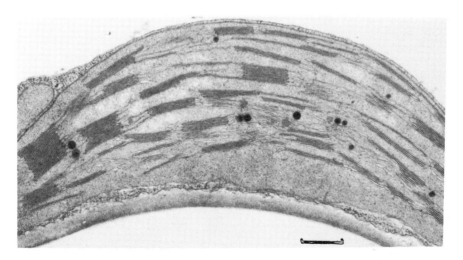

Figure 1.20 Electron micrograph of section through a spinach leaf chloroplast. Length of line 0.50 μm

Many other organelles can be 'identified' within eukaryotic cells including liposomes, peroxisomes, cytoplasmic filaments, microtubules, centrioles and vacuoles, each having a specific function within the cell.

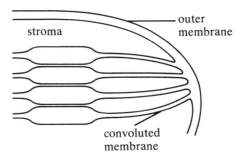

Figure 1.21 Sketch to show structure of the chloroplast membrane system

Questions

1. List the characteristics that distinguish living from non-living matter. Are the contents of a living cell at equilibrium with the environment?

2. The table gives the abundance (as percentage by mass) of the elements carbon, hydrogen, silicon, oxygen, nitrogen and phosphorus in living organisms and in the earth's crust.

Element	Organisms	Earth's Crust
Carbon	25.0	0.19
Hydrogen	49.0	0.22
Silicon	0.033	28.0
Oxygen	25.0	47.0
Nitrogen	0.27	0.1
Phosphorus	0.03	0.1

Plot these data in the form of a histogram. Comment on these data. In what form are carbon, hydrogen and oxygen present (chiefly) in living organisms?

3. What is the distinction between *bulk* nutrients and *trace* nutrients in animal nutrition?

4. List *five* properties of water which are of biological significance and explain the importance of any *two* of them.

5. The diagram below illustrates the structure of part of a body of water:

(a) Use this diagram to show your understanding of the distinction between covalent bonding and hydrogen bonding.
(b) How does hydrogen bonding help to explain the unique properties of water?
(c) Show how hydrogen bonding may occur between the molecules of two other compounds.
(d) The diagram illustrates part of the arrangement of amino acid residues in the polypeptide chain of a protein:

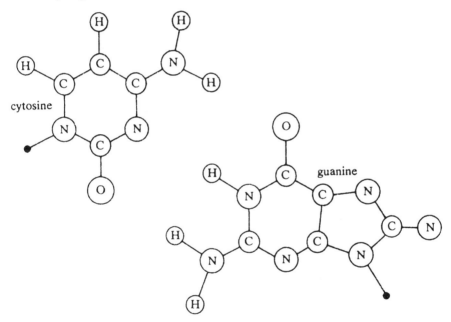

Use the diagram to show how hydrogen bonding may stabilize such a structure.

(e) The diagram below shows two compounds, cytosine and guanine, which are found in deoxyribonucleic acid (DNA). If these two compounds found themselves lying in this orientation with respect to each other, how might hydrogen bonds form?

6. Distinguish between the autotrophic and heterotrophic modes of nutrition. Show how heterotrophs, photoautotrophs and chemoautotrophs obtain:
   (a) raw materials for growth,
   (b) their energy supply.

7. (a) The instructions in a gardening book for making a general purpose fertilizer are as follows:
   sulphate of ammonia — 5 parts by weight
   superphosphate   — 5 parts by weight
   sulphate of potash   — 2 parts by weight
   (i) Which three elements is this fertilizer designed to provide?
   (ii) For each element state one function which it performs in plant cells.

(iii) This formula is for an 'inorganic' fertilizer which is usually applied to the soil during spring and early summer. Bone meal (a so-called 'organic' fertilizer) is usually applied in the autumn. Suggest why these application dates are different.
(b) With reference to man give:
   (i) three sources of calcium in the diet,
   (ii) three roles of calcium in the body,
   (iii) the mechanism controlling the concentration of calcium in the blood,
   (iv) two disorders which may occur if there is a deficiency of calcium in the diet.
[JMB]

8. The following diagram shows relationships between some nitrogen-containing compounds that are of importance in the nitrogen cycle.

$$urea \rightarrow X \rightleftharpoons NO_3 \rightarrow Y \rightarrow N_2$$

(a) Identify X and Y and write in on the diagram above. Write each of the following beside an appropriate arrow in the diagram:—urease, nitrate reductase, nitrifying bacteria.
(b) Place a ring around the compound that is most commonly absorbed by the roots of green plants.
(c) Name one process in the nitrogen cycle which is not included in this diagram.
[O & C]

9. List the roles of membranes within and at the surface of plant and animal cells. Select two of these activities and describe them in more detail. Relate, as far as possible, each activity to the structure of the membrane.
[O & C]

10. (a) Describe the structure and distribution of the membranes of cells.
(b) How do the different chemical components of membranes affect the properties of the membranes?
(c) Describe mechanisms by which extracellular material may enter a cell.
[JMB]

11. (a) Describe the uptake and metabolic importance of nitrogen in green plants.
(b) How is the supply of nitrogen maintained in a fertile loam?
[O & C]

12. Interpret each of the following observations. Include sufficient details of structure and function to explain your answer.
(a) The plasma membrane is selectively permeable: it is sometimes invaginated.
(b) In cells of animals and plants, vesicles pass from the Golgi bodies to the plasma membrane.
(c) Animal cells possess a centrosome.
(d) Root cap cells in plants contain amyloplasts in abundance.
(e) Large numbers of mitochondria surround the contractile vacuoles of a freshwater unicellular animal.
[JMB]

13. What is meant by the term 'organelle'?

In each of the electron micrographs *a–d*, identify the major organelle(s) present. For each of these state, in a sentence, its major biological function.

(a)   (b)

(c)   (d)

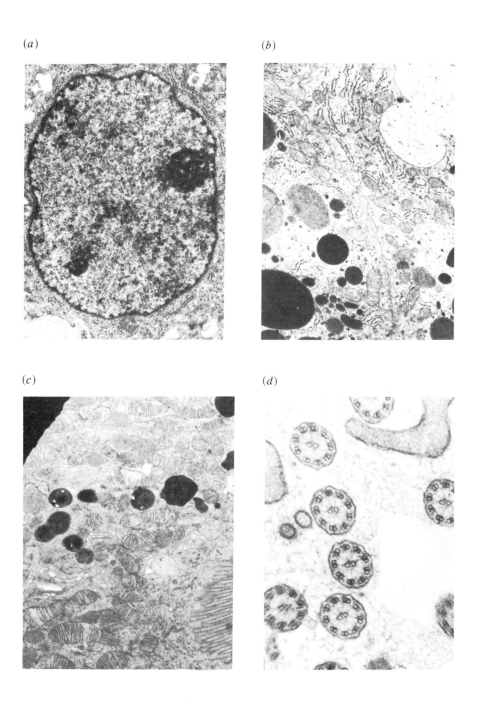

14. Describe how (*a*) microscope techniques and (*b*) cell fractionation techniques have helped our understanding of the structure of cells. Describe briefly the structure and state the function of any **four** of the following cell organelles: nucleus, mitochondria, endoplasmic reticulum, chloroplasts, cell membrane, Golgi apparatus.

15. The diagram shows a simplified carbon cycle.

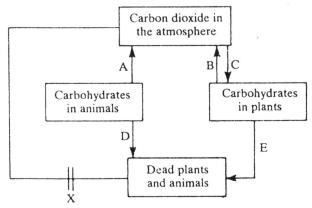

(*a*) Which pathway represents the conversion of inorganic carbon into organic form?

(*b*) If the cycle of events was blocked at X, what effect would this eventually have on (*i*) the composition of the soil, and (*ii*) the amount of carbon dioxide in the atmosphere?

[Scottish Higher]

Further Reading

The following *Scientific American* collections (published by W. H. Freeman & Co.) will be of interest in relation to this chapter: 'The Living Cell' (1965), and 'From Cell to Organism' (1967). In addition, the issue of *Scientific American* of September 1970 contains articles entitled 'The Carbon Cycle', 'The Nitrogen Cycle', 'Mineral Cycles', 'The Water Cycle', and several others.

Capaldi, R. A. (March 1974), 'A Dynamic Model of Cell Membranes', *Scientific American*, **230**, 26.

Cook, G. M. W. (1975), 'The Golgi Apparatus', *Oxford/Carolina Biology Reader*, **77**.

Fox, F. C. (February 1972), 'The Structure of Cell Membranes', *Scientific American*, **226**, 30.

Freedland, R. A., and Briggs, S. (1977), 'A Biochemical Approach to Nutrition', *Outline Studies in Biology*. (Chapman & Hall).

Frieden, E. (July 1972), 'The Chemical Elements of Life', *Scientific American*, **227**, 52.

Grimstone, A. V. (1977), 'The Electron Microscope in Biology', *Studies in Biology*, **9**. (Edward Arnold).

Horne, R. W. (1978), 'The Structure and Function of Viruses', *Studies in Biology*, **95**. (Edward Arnold).

Kaplan, M. M., and Koprowski, H. (January 1980), 'Rabies', *Scientific American*, **242**, 104.

Lockwood, A. P. M. (1979), 'The Membranes of Animal Cells', *Studies in Biology*, **27**. (Edward Arnold).

Lodish, H. F., and Rothman, J. E. (January 1979), 'The Assembly of Cell Membranes', *Scientific American*, **240**, 38.

2
Biological
Macromolecules

Summary

Macromolecules are essential to life processes, performing many vital tasks in the cell. All are built up from simple building blocks in condensation reactions whereby water is removed. Macromolecules can act as stores for times when there is no input of raw materials and energy from the environment. Starch and glycogen are carbohydrate stores. Fats also act as storage materials, although strictly speaking they are not macromolecules. Yet other types of macromolecule function as structural materials. These may be either polysaccharide, as in cellulose cell walls, or protein, as in tendons and cartilage. Yet other macromolecules, usually proteins, having precise and complex three-dimensional structures can interact with other molecules. Following interaction or 'recognition' of another molecule, a number of things may happen such as catalysis (enzymes), transport of gases in the blood (haemoglobin), regulation of life processes (hormones) and protection against infection (antibodies).

2.1 Introduction

All living organisms make and use certain classes of very large molecules called *macromolecules*. Some examples of macromolecules you may already have heard of are enzymes, haemoglobin, starch and DNA. Such molecules have special properties without which life would not be possible. Even the simplest cells contain several thousand different types of macromolecule. Molecular masses of macromolecules fall in the range from about 10 000 to millions. In contrast, the molecules normally dealt with by organic chemists have molecular masses in the range 50–500, although it must be admitted that twentieth-century life is highly dependent on synthetic polymers with very high molecular masses including plastics and man-made fibres.

Living organisms employ macromolecules in all their activities. The functions such molecules perform are very varied and include forming structural elements and food storage materials, transmitting and storing information, acting as highly efficient and specific catalysts, and having appropriate shapes for interacting with a variety of other molecules in life processes (see Table 2.1). The fact that they perform an enormous variety of functions means, of course, that macromolecules show an enormous range of structures. Despite this, we find that they are all built up according to certain common principles, and indeed almost all biological macromolecules belong to one of the three classes—*proteins, polysaccharides* or *nucleic acids*. One of the major triumphs of biochemistry is that the structures of very many macromolecules have been elucidated. Because we understand their chemical structures, we can explain their biological properties.

Table 2.1 Functions of some biological macromolecules

Function	Type of macromolecule	Example
Facilitating chemical reactions	Protein	Enzymes
Carrying oxygen in the blood	Protein	Haemoglobin
Chemical messenger	Protein	Hormones (e.g. insulin)
Food storage	Polysaccharide	Starch
Structure (plant cell wall)	Polysaccharide	Cellulose
Transmitting hereditary information	Nucleic acid	DNA

What are the common features of biological macromolecules?

Biological macromolecules are constructed by linking together a number of smaller molecules, or building blocks, to form long chains, with the exclusion of the elements of water. This class of reaction is called *condensation* (Figure 2.1). If all the building blocks are identical a *polymer* is formed. In contrast, if several types of building blocks are used and there is no 'repeating unit', then we get something very much more sophisticated and versatile, as we shall see shortly. In the same way that many different types of house may be built using half a dozen types of brick and tile, relatively few types of chemical building block serve in the construction of thousands of different types of macromolecules. Houses will show some of the properties of the bricks of which they are constructed, such as possessing a certain compression strength and being resistant to weathering, and so on. However, their interesting and useful *individual* features will result from the aggregate properties of the bricks—what we call *architecture*. The same is true of macromolecules. For a given class of macromolecule, the building blocks are drawn from a single, closely-related group of organic chemicals. Thus proteins are made up from building blocks called amino acids, and only about 20 different amino acids are used in the construction of literally tens of thousands of different and characteristic types of protein molecule in animals, plants and micro-organisms. Similarly, a dozen or two monosaccharide sugars are used in the construction of a wide variety of polysaccharide molecules, and a mere 4 nucleotide units suffice to build the nucleic acids (e.g. DNA) which can carry all the genetic information for the construction of a human being.

We can take the house analogy just one step further to help us understand the functions of macromolecules. A wall may be constructed from brick, composition block or from wood, and the choice of material will depend not only on the properties demanded of it, but also on economic factors such as cost and availability. This principle applies to living organisms too. Thus we find that availability of materials may dictate that some organisms use carbohydrates as a structural material while others use proteins.

EXAMPLE OF BASIC REACTION

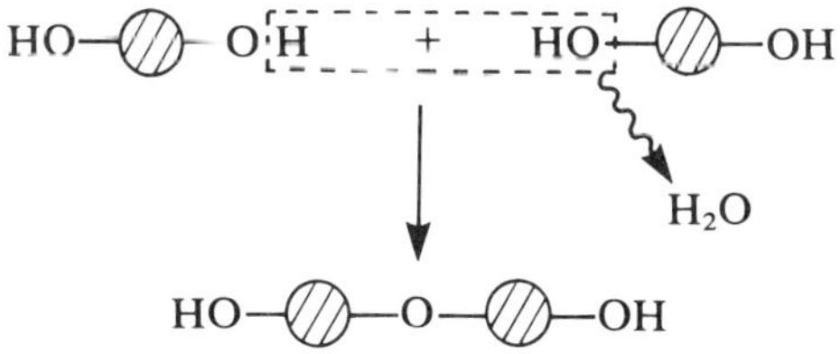

The groups that react do not necessarily have to be —OH

Joining the building blocks in a different order would give a molecule with different properties. If all the units are identical, or if there is a repeat, then the macromolecule produced is a *polymer*

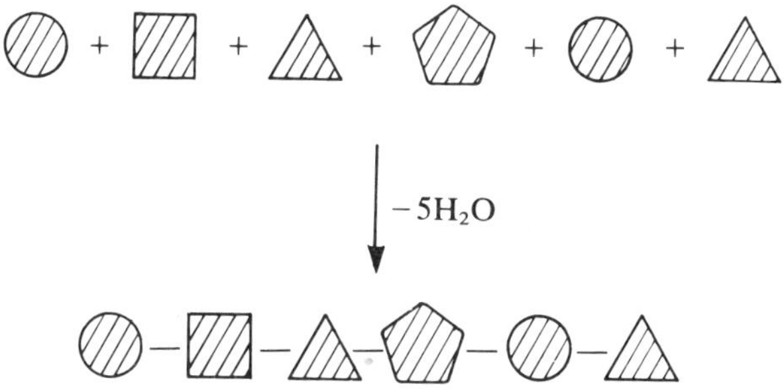

Figure 2.1 Formation of macromolecules by condensation reactions involving removal of water

In what follows we divide biological macromolecules into groups according to their functions. Their chief functions are as structural and storage materials, or to serve in 'information and control situations'. Included in the latter category are the enzymes whose function it is to catalyze or control the rates of chemical reactions. A simple polymer with a repeating unit is often sufficient for the first type of function, but a much more complicated sequence in the macromolecule is necessary for the second type of function. In the appropriate places we shall consider the nature of the building blocks and how they are joined together. However, we shall try to keep in mind at all times that it is the desire to understand *function* that leads the biochemist to study the chemistry of living systems.

2.2 Storage of food

The maintenance of life requires a continuous input of energy. However, living organisms do not necessarily have access to supplies of energy from the environment at all times. For example, photosynthesis in plants may only proceed when the sun is shining and predatory animals may be forced to survive long periods between meals. In order to reconcile these two opposing situations, most living organisms have evolved ways of storing food materials.

We shall see in Chapter 3 how energy is released by the oxidation of carbon compounds to carbon dioxide. For the present it is enough to note that both food materials and storage compounds are in a *more reduced* form than carbon dioxide. In fact, almost all organic carbon compounds fulfil this requirement, but in practice only a few types of compound are stored. Why is this?

What sort of compounds make good storage materials?

The ideal storage material would be something that is reasonably compact and inert, but which can be mobilized quickly when food materials are unavailable from the environment. The two types of material that have been 'chosen' for this role during the course of evolution are *polysaccharides* and *lipids*. Although proteins and nucleic acids are used as foods by heterotrophs, they rarely serve as stores of 'reduced carbon'. Probably this is because they are potentially capable of interacting with other molecules and it could therefore raise problems to store them. Protein is occasionally used as a transferable food material. For example, the protein *casein* in milk is made by the mother and is passed to the offspring where it is broken down to amino acids for growth. In this example, however, the amino acids are used as building blocks rather than as energy-providing materials.

Polysaccharide stores

Polysaccharides are polymers of sugar molecules, and 'sugar' is a colloquial term used to describe certain members of the class of compounds called *carbohydrates*, empirical formula $(CH_2O)_n$. Carbohydrates are important in many life processes, and sugars are familiar as constituents of many foodstuffs including glucose and fructose in honey, and sucrose in cane and beet sugar. Many of the reactions going on in living cells are concerned with the degradation of sugar molecules in order to release the energy that the molecules contain, i.e.,

$$(CH_2O)_n \xrightarrow{\;O_2\;} CO_2 + H_2O + energy$$

Carbohydrates also form structural materials such as cellulose (see p. 77) and the capsules of bacterial cells. Generally speaking, sugar molecules are very soluble in water and would be difficult to store in high concentrations in the aqueous environment of cells. There is the difficulty of holding the molecules all in one place and high concentrations of small molecules produce high osmotic pressures resulting in water being drawn into the cell. The answer to this problem is to link single sugar molecules together to form polymers which are relatively insoluble and which exert much lower osmotic pressures. We must pause here to study some aspects of the chemical structures of sugars in order to see later how they function as sources of energy and how they can be linked into polymers.

The structure of carbohydrates

Sugars, or carbohydrates, are characterized by the possession of aldehyde (—CHO) or ketone ($>C=O$) groupings in addition to two or more hydroxyl (—OH) groups. The simplest carbohydrate is glyceraldehyde,

$CH_2OH-CHOH-CHO$, and other members of the group are molecules containing more $-CHOH-$ groupings, e.g. $CH_2OH-(CHOH)_4-CHO$. If we try to draw the structure of glyceraldehyde in three dimensions or build a model of it, we find that it can be done in two ways (Box 2.1) because of the tetrahedral arrangement of the four chemical groups which bond to the carbon atom. The two forms of glyceraldehyde are distinct and one form can *only* be converted into the other in a reaction involving the breaking and reforming of covalent bonds. The two forms are actually so similar in chemical properties that they are very difficult to distinguish. Chemists have found that solutions of these compounds rotate the plane of plane-polarized light in opposite directions, and the two are therefore called 'optical isomers'. Biological organisms are not concerned with optical rotation, but they *do* deal with three-dimensional molecules and have no difficulty in recognizing these two as quite distinct structures (Box 2.1). Many biological molecules exist as optical isomers and this includes not only almost all carbohydrates, but also amino acids. In general, biological systems deal with only one of two possible isomers. (The reasons for this will become clearer later when we consider how molecules 'recognize' each other.)

Box 2.1 *Structure of glyceraldehyde and optical activity*

Glyceraldehyde is the aldehyde derived from the trihydric alcohol glycerol:

CH_2OH	CHO
$CHOH$	$CHOH$
CH_2OH	CH_2OH
glycerol	glyceraldehyde

When we come to build a three-dimensional model of glyceraldehyde we find that we can do it in two ways.

These two are very similar in chemical and physical properties but they are not identical. In fact, any compound which has four different chemical groupings attached to a single carbon atom will show this type of isomerism. Because the different isomers rotate the plane of polarized light in opposite directions they can be distinguished and because of this property they are called *optical isomers*.

$$R_2-\overset{\overset{\displaystyle R_1}{|}}{\underset{\underset{\displaystyle R_3}{|}}{\text{C}}}-R_4$$

The isomer that rotates the plane of polarized light to the right is called the dextrorotatory form and the other the laevorotatory form. The extent of optical rotation is measured in an instrument called a *polarimeter*. Living organisms have no difficulty in distinguishing the two three-dimensional structures. The structures are in fact mirror images of each other like the two hands of a pair of gloves.

In order to represent such three-dimensional structures on two-dimensional paper certain rules must be followed. These are that the most reactive grouping or carbon atom 1 must be written at the 'top', and that the two-dimensional representation must not be removed from the plane of the paper, although it may be 'slid around'. On this basis we can write the above two forms of glyceraldehyde:

$$\overset{1}{\text{CHO}} \qquad\qquad \overset{1}{\text{CHO}}$$
$$H-\overset{2}{C}-OH \qquad HO-\overset{2}{C}-H$$
$$\overset{3}{\text{CH}_2\text{OH}} \qquad\qquad \overset{3}{\text{CH}_2\text{OH}}$$

There is no way in which these can be 'matched' if they are kept within the plane of the paper. With reference to the three-dimensional structure: bonds drawn horizontally point 'up' from the plane of the paper and bonds drawn vertically point 'down'.

Box 2.2 Carbohydrates

Carbohydrates are polyhydroxy aldehydes or ketones. The simplest ones are:

$$\text{CHO} \qquad\qquad \text{CH}_2\text{OH}$$
$$H-C-OH \qquad C=O$$
$$\text{CH}_2\text{OH} \qquad\qquad \text{CH}_2\text{OH}$$
glyceraldehyde   dihydroxyacetone

Adding —CHOH— units gives a series of possible compounds thus:

$$
\begin{array}{cccc}
& & \text{CHO} & \text{CHO} \\
& \text{CHO} & H-C-OH & H-C-OH \\
\text{CHO} & H-C-OH & H-C-OH & H-C-OH \\
H-C-OH & H-C-OH & H-C-OH & H-C-OH \\
\text{CH}_2\text{OH} & \text{CH}_2\text{OH} & \text{CH}_2\text{OH} & H-C-OH \\
& & & \text{CH}_2\text{OH}
\end{array}
$$

The five-carbon compound above is ribose. The most common six-membered compound is glucose (not shown). Because each of the carbon atoms, except the first and the last, is asymmetric, a number of isomers are possible in which mirror-image forms of *each* of these carbon atoms are present. Only a very small number of all the possible isomers actually occur in nature.

Glucose has the straight chain structure shown below but it tends to cyclize:

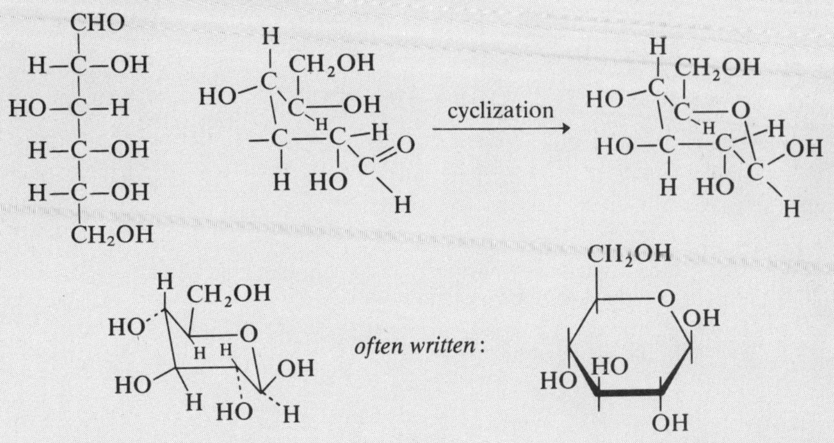

CHO
H—C—OH
HO—C—H
H—C—OH
H—C—OH
CH$_2$OH

cyclization →

This cyclization can happen in two different ways to give two different isomers:

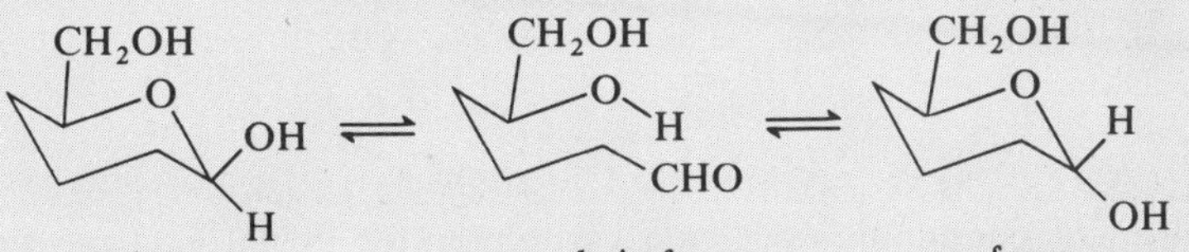

β-form   open-chain form   α-form

In solution these are present in equilibrium *through* the open chain form (which is only present in tiny amounts). Although the so-called α- and β-glucose themselves are very similar in properties, it makes a great deal of difference whether sugar polymers are formed through α links or β links (see p. 77).

Ring numbering. In order to identify the carbon atoms in sugars there is a numbering system. The aldehyde carbon is always called carbon 1. Thus glucose is numbered:

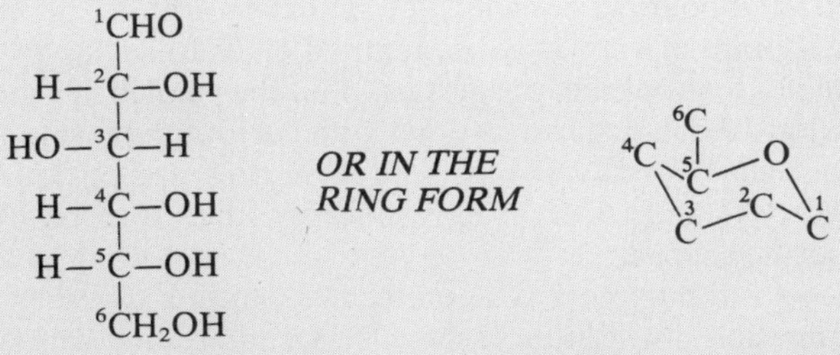

^{1}CHO
H—^{2}C—OH
HO—^{3}C—H
H—^{4}C—OH
H—^{5}C—OH
^{6}CH$_2$OH

OR IN THE
RING FORM

Optical activity of carbohydrates. It is impossible to predict from the structure of a carbohydrate in which direction the plane of polarized light will be rotated, i.e. whether it is laevo- or dextrorotatory. A convention has therefore been established which relates the structure of all carbohydrates to the two optical isomers of glyceraldehyde (see Box 2.1). This convention states that those carbohydrates in which the last optically-active carbon atom has the same configuration as in dextro-glyceraldehyde will be said to belong to the so-called 'D-series'. The others will be said to belong to the 'L-series'. Unfortunately, *this says nothing about the direction of rotation of the plane of polarized light*—but we have said this is of no interest to biological organisms, which 'look at' three-dimensional structures not optical rotation. Thus:

$$
\begin{array}{ccc}
\text{CHO} & \text{CHO} & \text{CHO} \\
\text{H–C–OH} & \text{H–C–OH} & \text{H–C–OH} \\
\text{CH}_2\text{OH} & \text{H–C–OH} & \text{HO–C–H} \\
 & \text{CH}_2\text{OH} & \text{H–C–OH} \\
 & & \text{H–C–OH} \\
 & & \text{CH}_2\text{OH}
\end{array}
$$

note that this last carbon atom is not optically active

all belong to the D-series. A curious consequence of this is that D-glucose is dextrorotatory while D-fructose is laevorotatory

$$
\begin{array}{cc}
\text{CHO} & \text{CH}_2\text{OH} \\
\text{H–C–OH} & \text{C=O} \\
\text{HO–C–H} & \text{HO–C–H} \\
\text{H–C–OH} & \text{H–C–OH} \\
\text{H–C–OH} & \text{H–C–OH} \\
\text{CH}_2\text{OH} & \text{CH}_2\text{OH}
\end{array}
$$

D-glucose   D-fructose

The most commonly-occurring carbohydrates in nature have five or six carbon atoms and are referred to as pentose and hexose sugars, respectively (Box 2.2). The most frequently encountered are glucose (six-carbon, aldehyde), fructose (six-carbon, ketone) and ribose (five-carbon, aldehyde). In solution these exist almost entirely in the 'ring' forms (Figure 2.2) with an oxygen atom in the ring, and with the aldehyde or ketone properties largely hidden. Glucose, for example, is not oxidized to the corresponding acid in the presence of atmospheric oxygen to any appreciable extent as would be expected of an aliphatic aldehyde. The main features of the molecules from the *biological* point of view may be summed up as follows:
(1) They have hydrophilic properties, i.e. the molecules possess several hydroxyl groups which can interact with surrounding water molecules.
(2) They have characteristic shapes. The sugar rings are 'puckered', and bristle with hydroxyl and other groups, and can therefore be 'recognized' by their specific shapes.
(3) They can form polymers. The molecules can link together in chains with the elimination of water molecules, i.e. condensation can take place.
In addition, it is possible to form many derivatives of carbohydrates in which the hydroxyl groups are replaced by other groupings (Figure 2.3). This results in an even larger variety of compounds with an enormous diversity of shapes and properties.

Polysaccharide structure

Polymerization can occur with the elimination of a molecule of water as shown in Figure 2.4 although the reaction is more complicated than this in living cells (see p. 162). The reason for this complication is to make sure the desired

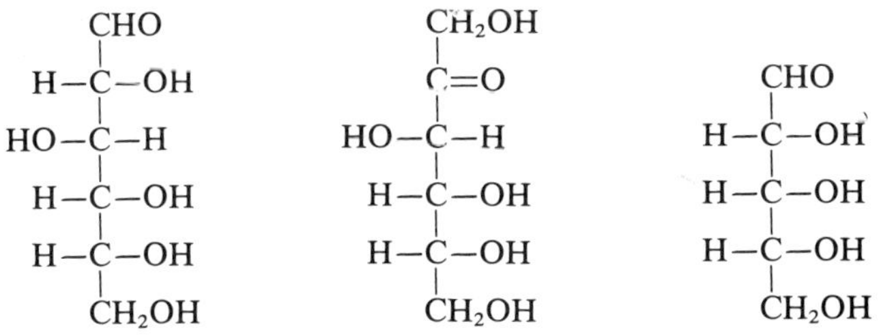

glucose fructose ribose

The open chain forms of glucose, fructose and ribose are represented in two dimensions, using the convention (Box 2.2), as above.

In solution these exist largely in the form of ring structures. Glucose has a strong tendency to form a six-membered ring whereas the other two sugars tend to form five-membered rings. For convenience the ring structures are often simplified by omitting the carbon and hydrogen atoms.

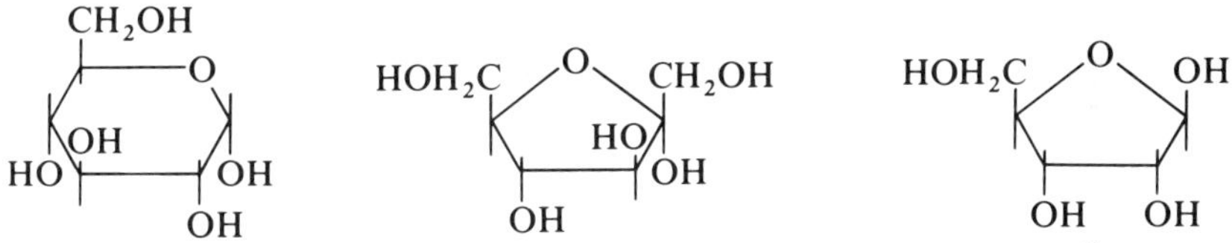

glucose fructose ribose

Sometimes these are shown as perspective drawings:

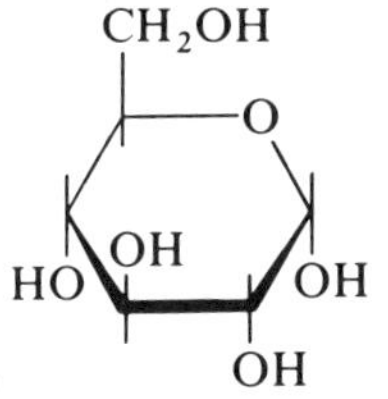

but we should remember that the rings are puckered

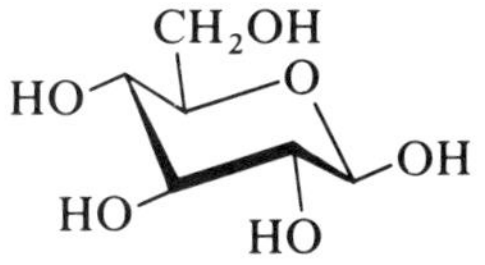

Figure 2.2 Structures of glucose, fructose and ribose

polysaccharide is formed. Thus, for any six-carbon sugar there are potentially five hydroxyl groups that could form a link with another sugar. Since the mode of linking determines the physical properties of the polymer, it is important to ensure that the correct one is used.

The main storage polysaccharide in many plants is starch, and in animals, glycogen (Figure 2.5). These are both large molecules consisting of hundreds of glucose molecules linked together. Most of the links are between carbon 1 of one glucose residue and carbon 4 of the next, but there are branch points in which the links are $\alpha 1 \rightarrow 6$. The resulting highly-branched structure tends to be more or less spherical, although one component of starch, amylose, has no branch points and forms a helical structure (Figure 2.5).

A food storage molecule must be able to release its small molecular components for use when food is no longer being taken from the environment. Starch and glycogen can do this when acted upon by the appropriate biological catalysts or enzymes (see p. 142). But, also, they fit nicely into the cell, not occupying too much space, by forming granules, nor upsetting the osmotic

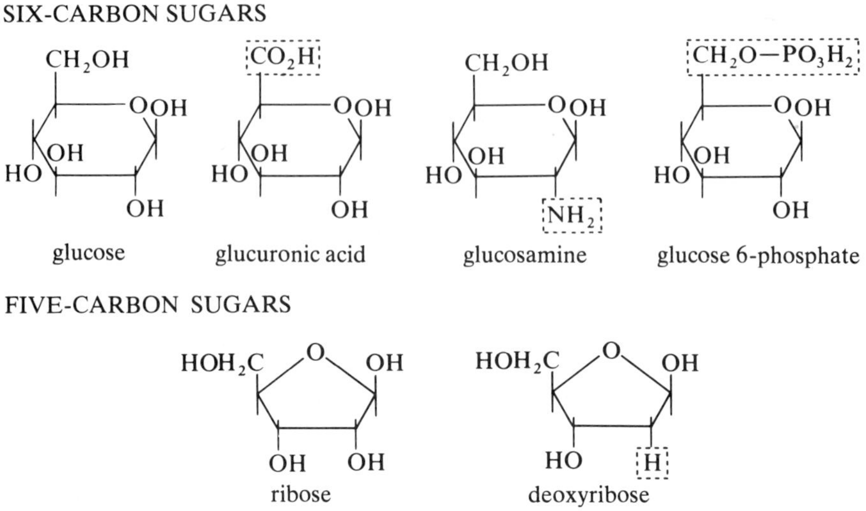
SIX-CARBON SUGARS
CH_2OH
OOH
OH
HO
OH
glucose
CO_2H
glucuronic acid
glucosamine
NH_2
$CH_2O-PO_3H_2$
glucose 6-phosphate
FIVE-CARBON SUGARS
HOH_2C
O
ribose
deoxyribose
H

Figure 2.3 Some carbohydrate derivatives that occur naturally

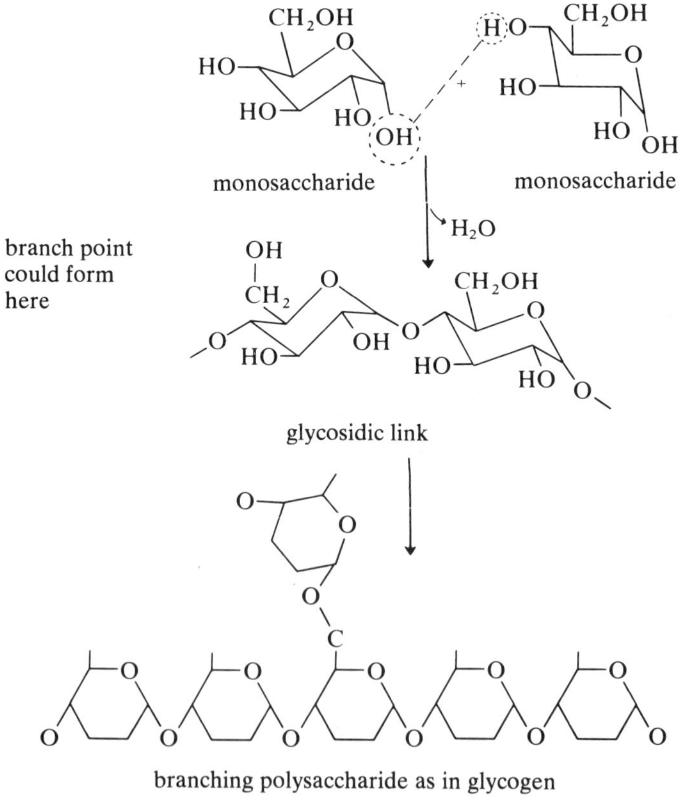
CH_2OH
HO
OH
monosaccharide
monosaccharide
H_2O
branch point
could form
here
CH_2
O
glycosidic link
C
branching polysaccharide as in glycogen

Figure 2.4 Formation of polysaccharides

Starch contains two types of polymer, the straight chain *amylose*, shown above having only $\alpha 1 \rightarrow 4$ links, and the branched chain *amylopectin*, which has a structure similar to that of glycogen, shown below.

Glycogen, the storage polysaccharide of vertebrates, consists of glucose molecules linked $\alpha 1 \rightarrow 4$ with $\alpha 1 \rightarrow 6$ branch points to form bush-like molecules

Figure 2.5 The storage polysaccharides starch and glycogen. Both are polymers of glucose

balance of the cell. Glucose polymers such as starch and glycogen may be equivalent to a 3 M glucose solution compressed into a single granule which is osmotically inert. The branched, porous structure allows attack by enzymes when appropriate.

Disaccharides for sugar transport

Sugars are often transported round organisms, sometimes as simple sugars (monosaccharides), sometimes as disaccharides in which two monosaccharides are linked together in the same fashion as in polysaccharides (compare Figures 2.5 and 2.6). Higher animals invariably transport glucose in the bloodstream, but milk contains the disaccharide lactose or milk sugar (Figure 2.6) in addition to the protein mentioned above. Plants usually transport sugars about their tissues as concentrated solutions of another disaccharide, sucrose, and this compound has, of course, immense commercial importance. Yet another disaccharide, trehalose, is important in insects and some fungi (Figure 2.6).

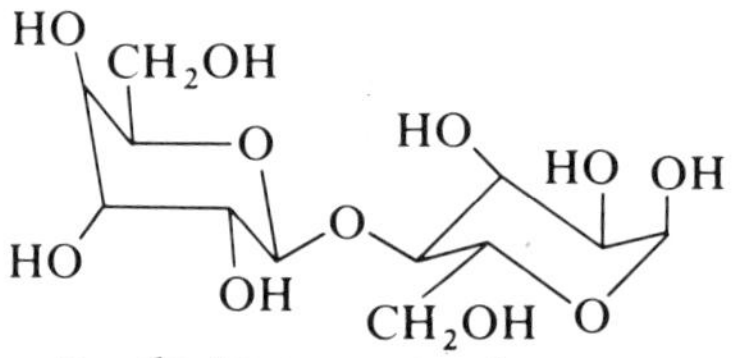

D-galactose   D-glucose
LACTOSE

Function: transport of carbohydrate from mother to infant in milk

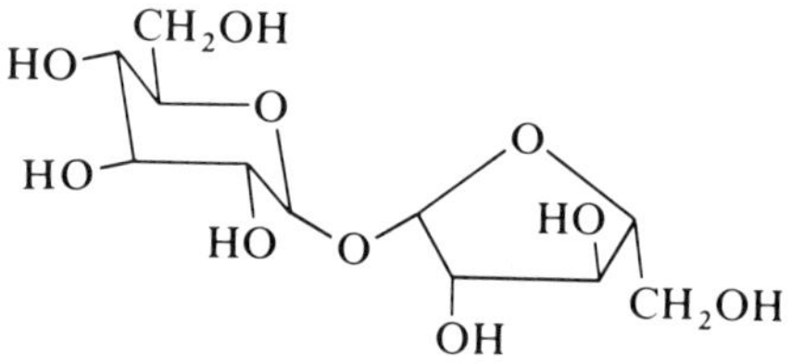

D-glucose   D-fructose
SUCROSE

Function: transport of carbohydrate from one part of a plant to another

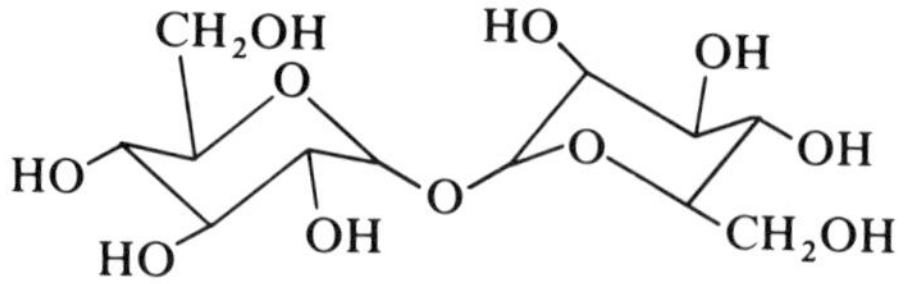

D-glucose   D-glucose
TREHALOSE

Function: carbohydrate transport in insects

Figure 2.6 Disaccharides serve a variety of purposes

Fat stores

Although a great deal of energy is stored as polysaccharide by both animals (glycogen) and plants (starch), an alternative storage material is fat. Fats belong to the class of compounds called 'lipids'. They are not macromolecules, but because of their hydrophobic (i.e. water-hating) nature the molecules tend to aggregate to form globules. They are therefore good storage materials because the globules represent compact, relatively inert, stores of energy. It is convenient to deal with lipids at this point whilst we are thinking about storage materials. Our approach throughout is to observe a phenomenon or function—here storage—and then seek a chemical explanation for it. In this case there happen to be two chemical ways of storing energy which are quite distinct, namely the use of polysaccharides and the use of lipids. Most organisms use both.

What is lipid?

We are familiar with fats in the kitchen such as lard, butter and dripping, and these are all 'lipids'. Natural oils such as olive oil have an almost identical structure to these, but are called oils because they are liquid at room temperature rather than solid. If a fat such as lard, or an oil such as olive oil, is boiled up with alkali (NaOH) a process known as 'saponification', or soap formation, takes place. The fats and oils are esters of the alcohol, glycerol, and fatty acids, and hydrolysis produces a mixture of glycerol and the sodium salts of the fatty acids. This mixture has detergent properties and we call it a 'soap'. The ester compounds are called triglycerides (Box 2.3) and are neutral because the acid function is concealed; you will note that these are formed by a condensation reaction. (Other lipids such as some of those that occur in membranes are not neutral.)

The different properties of fats and oils arise because of the different fatty acids of which they are constituted. These fatty acids may have short or long carbon chains, and may be saturated or unsaturated (Box 2.3). Triglycerides containing fatty acids with short chains and those with unsaturated fatty acids (i.e. having double bonds) will have lower melting points than those whose fatty acids have long chains and a high degree of saturation.

Box 2.3 Fats

Naturally-occurring fats and oils are triglyceride esters of glycerol with fatty acids. A fatty acid is simply a compound with the structure:

$CH_3(CH_2)_nCO_2H$ REPRESENTATION: $\wedge\!\wedge\!\wedge\!\wedge\!\wedge\!\wedge\!\wedge\!\wedge\!\wedge$ CO_2H

Where n may be anything from 2 to 20 or more and where, in addition, there may be double bonds present. Hydrolysis of neutral triglycerides yields 1 molecule of glycerol and 3 molecules of fatty acids:

$$CH_3(CH_2)_nCOO \begin{array}{l} CH_2OCO(CH_2)_nCH_3 \\ | \\ CH \\ | \\ CH_2OCO(CH_2)_nCH_3 \end{array} \rightarrow \begin{array}{l} CH_2OH \\ | \\ CHOH \\ | \\ CH_2OH \end{array} + 3CH_3(CH_2)_nCO_2H$$

neutral triglyceride   glycerol   fatty acids

The character of fatty acids, and hence of triglycerides, is determined by the length of the carbon chain and by the number of double bonds:

$CH_3(CH_2)_{16}CO_2H$ $CH_3(CH_2)_2CO_2H$ $CH_3(CH_2)_7CH{=}CH(CH_2)_7CO_2H$
stearic acid from   butyric acid from   oleic acid from
beef fat   butter   olive oil

long chain, saturated   *short chain*   *long chain, unsaturated*

Short-chain, and unsaturated fatty acids, and their derivatives tend to have lower melting points than long chain, and saturated acids. In living organisms the triglycerides occur in the liquid state. It is therefore important that an organism makes the right kind of fat: fat from an animal with a body temperature of 37 °C would not suit a fish living in Arctic waters!

In the ionized form, free fatty acids, $\wedge\wedge\wedge\wedge\wedge\wedge\wedge$ CO_2^-, have detergent properties because one end of the molecule (CO_2^-) is water-soluble or hydrophilic, while the other end is hydrophobic. Such molecules form micelles and bilayers, and can also disrupt organized structures such as membranes.

Properties of lipid stores

Neutral triglycerides are almost completely insoluble in water, and collections of such molecules tend to come together and segregate from the aqueous phase forming a spherical droplet. The fatty acids themselves are a good store of energy because they represent carbon in a highly-reduced form. The non-covalent aggregation of the triglyceride molecules is equivalent to the covalent linking of glucose molecules to form a polymer such as starch.

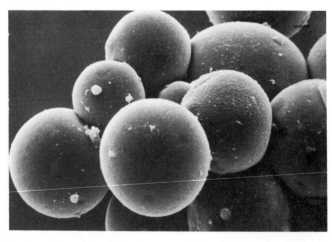

Figure 2.7 Isolated fat cells appear to be spherical except for a small protrusion of the nucleus. The cytoplasm forms a very thin layer over the surface of the central lipid droplet. Analysis has showed that the lipid content is 70–80% of the wet weight of the cells. These cells were isolated by treating rat epididymal adipose tissue with a collagen-digesting enzyme. The isolated cells were then fixed, dehydrated and sputtered with gold, and the picture taken in the scanning electron microscope. The diameter of the cell at the front is about 0.05 mm. (*Photograph kindly supplied by Nils Östen Nilsson, University of Lund, Sweden*)

This is how lipid is stored, and cells of adipose tissue are very often engorged with large droplets of lipid (Figure 2.7). There are additional advantages in having tissues composed of large numbers of such lipid-containing cells. The lipid is a good heat insulator, and the adipose tissue can also act as a mechanical protection under the skin and around vital organs. Mammals have exploited this property to the full. Diving mammals use the heat-retaining properties, and there is the additional bonus that fat, being less dense than water, gives buoyancy.

Although fat is a very efficient energy store it has one disadvantage, that its energy content cannot be released in the absence of oxygen. (The mechanisms of this process will become clear in Chapter 3.) Normally this would not matter in an aerobic organism, but in rapidly-contracting voluntary muscle of vertebrates it is crucial because blood is squeezed out by the contractions and no oxygen can be supplied. Therefore such tissue must obtain an energy supply anaerobically—and as it turns out, only carbohydrates can supply energy anaerobically. Hence the requirement for two types of energy store.

2.3 Information, interaction and control

Storage macromolecules are, comparatively speaking, very simple in structure. They could be described by a sequence ...A.A.A.A.A.A.A..., perhaps with some branching. It is clear that if we had a molecule with a different and non-repeating structure (not strictly a polymer) such as ...A.B.A.A.D.C.B.B.A.C... it could be used to store information in much the same way as the punched tape used with computers. This is indeed how living organisms store information in nucleic acids: however, we shall defer discussion of this topic until Chapter 5 when we deal with biochemistry and its relationship to genetics.

Another possible situation is when the units in the above non-repeating structure have different chemical properties—some hydrophilic, some hydrophobic, some negatively charged, some positively charged, etc. A folding of the molecule could produce a three-dimensional structure which could 'match', or form a complementary shape to, another molecule. Such a structure would *interact* with other molecules, and such interactions could be useful in controlling reactions and other activities in cells. This is how *proteins* work as enzymes, hormones, antibodies and carriers.

Weak bonding

When macromolecules coil up and when they interact with other molecules the interactions depend not on covalent bonds, but on so-called 'weak bonds' which can readily be broken and remade (Table 2.2). The sort of energies involved are very much smaller than those relating to covalent bonds. Collections of such bonds in a macromolecule will tend to be in a dynamic state, continually breaking and reforming with fluctuations in the thermal energy.

We have already mentioned hydrogen bonds (p. 9) in Chapter 1 and have considered the idea of hydrophobic interactions between non-polar groupings resulting in the exclusion of water molecules such as in the formation of lipid globules. Van der Waal's forces or bonds occur only over short distances between atoms. They occur because slight fluctuations in the electron clouds around certain chemical groupings result in the formation of transient dipoles, i.e. regions with a separation of positive and negative charge. One dipole will attract another with the opposite orientation, like two bar magnets placed with unlike poles together.

Table 2.2 Bond energies of types of bond occurring in macromolecules

Type of bond	Energy (kJ mol^{-1})
Covalent	200–400
Hydrogen	up to 20
Van der Waal's	up to about 4
'Hydrophobic'	4–8
Ionic	up to 4

'Ionic bonding' is the electrostatic interaction between oppositely-charged ions such as $-CO_2^-$ and $^+H_3N-$. None of these bonds by themselves are very strong or significant in maintaining a molecule in a particular three-dimensional structure. However, taken together, several hundred of them act to hold macromolecules in characteristic shapes. Usually the performance of biological function depends upon macromolecules having definite shapes. We shall now see how this effect is achieved by studying the proteins, and especially a particular class of globular, soluble proteins, the enzymes.

2.4 The proteins

The proteins represent a major class of macromolecule found in all living organisms. There are literally hundreds of thousands of different types of protein and an individual cell may contain several thousand different ones. Proteins play a number of roles within the cell. Those playing a structural role will be considered at the end of this chapter (p. 77). Others, the so-called 'globular or soluble proteins' will be considered now. However, all the proteins are constructed in the same fashion.

Proteins are built up by linking together chemical units called amino acids. The linking reaction is, of course, a condensation (Figure 2.8) and the bond formed is called a *peptide bond*. About twenty types of amino acids are typically found in living systems—the same twenty regardless of the cell type or organism. Just as linking together the twenty-six letters of our alphabet produces an enormous range of words, sentences, and eventually books, linking together amino acids produces an extremely versatile class of macromolecules.

Are proteins polymers?

Proteins are typically unbranched chains consisting of several hundred amino acid units linked by peptide bonds. Such a structure is called a polypeptide chain, but it is not really a polymer. It has a fixed composition and molecular mass, and it contains information by virtue of the specific *sequence* of the different types of amino acid in the chain. The 20 amino acids typically found in proteins are shown in Table 2.3.

All of the naturally-occurring amino acids are α-amino acids and have the general structure:

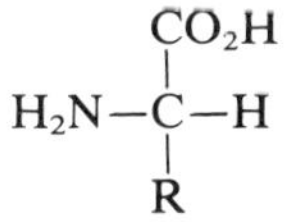

$$CO_2H$$
$$H_2N-\overset{|}{\underset{|}{C}}-H$$
$$R$$

They are called 'α-amino acids' because both of the functional groups are attached to the same or α carbon atom. All of the naturally-occurring amino acids, except glycine, are optically active, and only the L-series amino acids occur in proteins.

Amino acids exist in different forms depending upon the pH of the medium in which they are dissolved:

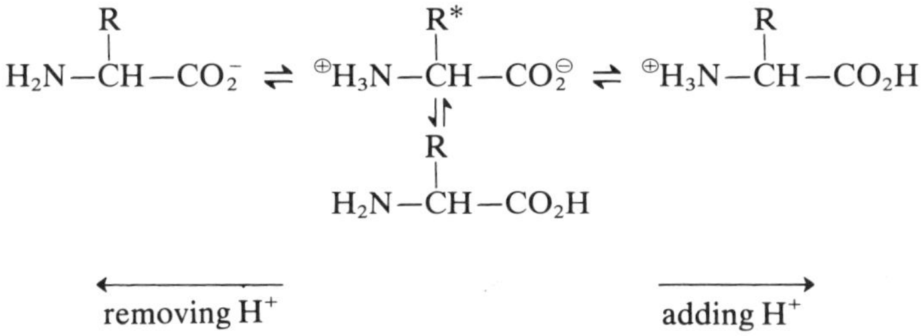

$$\underset{R}{H_2N-CH-CO_2^-} \rightleftharpoons \overset{\oplus}{H_3}N-\underset{R^*}{CH}-CO_2^{\ominus} \rightleftharpoons \overset{\oplus}{H_3}N-\underset{R}{CH}-CO_2H$$

$$\updownarrow$$

$$H_2N-\underset{R}{CH}-CO_2H$$

$\xleftarrow{\hspace{2cm}}$ \qquad\qquad $\xrightarrow{\hspace{2cm}}$

removing H^+ \qquad\qquad\qquad adding H^+

*This form, the so-called zwitterion form, probably represents the 'neutral', uncharged form, rather than the undissociated form shown below it.

Amino acids, peptides and polypeptides tend to be charged because of their $-NH_2$ and $-CO_2H$ groups ionizing and also because of charges on the 'R' groups (see Table 2.3). There will of course be one pH value at which the molecule will have a zero net charge: this is called the 'iso-electric point'. At any other pH the molecule will have a charge and because of this will move in an electric field. Positively-charged molecules will move towards a negative electrode and *vice versa*. This movement in an electric field is called 'electrophoresis' and is frequently used as a technique for separating mixtures of amino acids, peptides or proteins into their components. Typically a piece of filter-paper is soaked in a solution of given pH. A spot containing the mixture to be separated is placed on the paper and then a direct current is applied across the paper. The various components move either to the cathode or to the anode, and at different rates, depending upon their charges (see Box 5.8).

Amino acids can react together with the elimination of water to form a peptide. Long strings of amino acids linked in this way are called polypeptides.

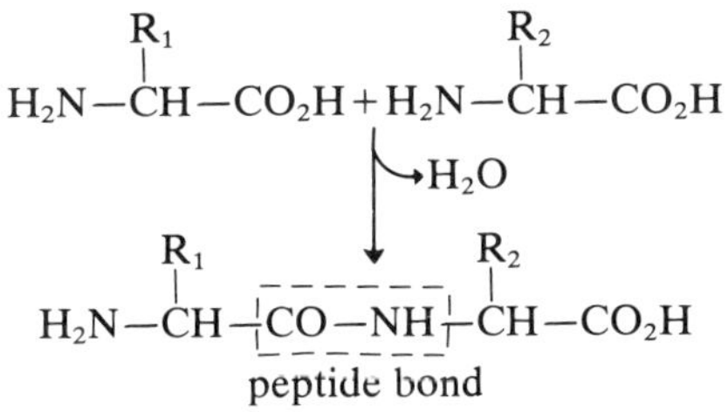

$$H_2N-\underset{R_1}{CH}-CO_2H + H_2N-\underset{R_2}{CH}-CO_2H$$

$$\searrow H_2O$$

$$H_2N-\underset{R_1}{CH}-CO-NH-\underset{R_2}{CH}-CO_2H$$

peptide bond

Because of the arrangement of the atoms and electrons, the peptide bond is planar: this limits the possible orientation of one amino acid in a polypeptide chain compared with the next one

Figure 2.8 Amino acids and peptide bonds

Table 2.3 The amino acids found in proteins shown both as chemical structures and space-filling models

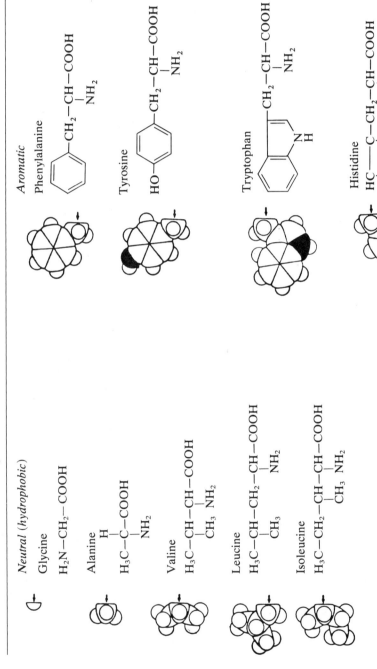

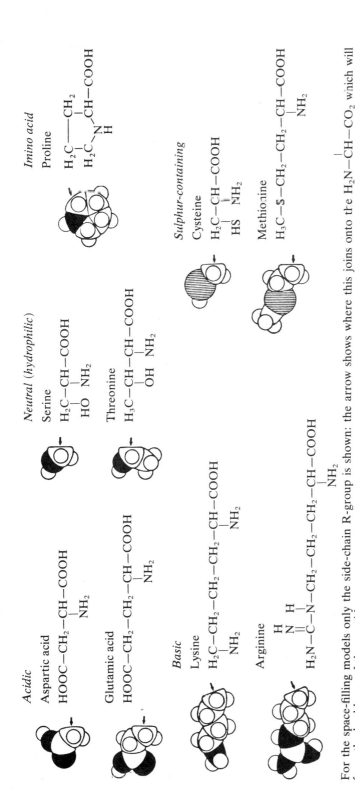

Acidic

Aspartic acid

$$\text{HOOC—CH}_2\text{—CH—COOH}$$
$$\text{NH}_2$$

Glutamic acid

$$\text{HOOC—CH}_2\text{—CH}_2\text{—CH—COOH}$$
$$\text{NH}_2$$

Basic

Lysine

$$\text{H}_2\text{C—CH}_2\text{—CH}_2\text{—CH}_2\text{—CH—COOH}$$
$$\text{NH}_2 \qquad\qquad\qquad \text{NH}_2$$

Arginine

$$\text{H}_2\text{N—C—N—CH}_2\text{—CH}_2\text{—CH}_2\text{—CH—COOH}$$
$$\text{N} \quad \text{H} \qquad\qquad\qquad\qquad \text{NH}_2$$

Neutral (hydrophilic)

Serine

$$\text{H}_2\text{C—CH—COOH}$$
$$\text{HO} \quad \text{NH}_2$$

Threonine

$$\text{H}_3\text{C—CH—CH—COOH}$$
$$\text{OH} \quad \text{NH}_2$$

Imino acid

Proline

$$\text{H}_2\text{C—CH}_2$$
$$\text{H}_2\text{C} \quad \text{CH—COOH}$$
$$\text{N}$$
$$\text{H}$$

Sulphur-containing

Cysteine

$$\text{H}_2\text{C—CH—COOH}$$
$$\text{HS} \quad \text{NH}_2$$

Methionine

$$\text{H}_3\text{C—S—CH}_2\text{—CH}_2\text{—CH—COOH}$$
$$\text{NH}_2$$

For the space-filling models only the side-chain R-group is shown: the arrow shows where this joins onto the $\text{H}_2\text{N—CH—CO}_2$ which will form the backbone of the peptide.

The chemical structures are shown in their unionized form, i.e. $-CO_2H$ instead of $-CO_2^-$, etc. Note that the amides, glutamine and asparagine, of the acidic amino acids, glutamic and aspartic acid, also occur in proteins with the terminal $-CONH_2$ instead of $-CO_2H$. Three-letter abbreviations are used when writing peptide and protein structures. Normally the first three letters of the name are used, e.g. *ala = alanine*, except in the case of tryptophan which is abbreviated '*trp*'. Glutamine and asparagine are written *gln* and *asn*, in order to distinguish them from the acids, glutamic acid (*glu*) and aspartic acid (*asp*). More recently a single letter code has been invented to make writing peptide structures three times as quick. It is not, however, as logical (useful though it is to those writing peptide structures every day): for example, lysine is designated 'K' to distinguish it from leucine 'L'.

In proteins there is a repeating backbone of peptide bonds thus:

$$\ldots -CO-NH-\overset{\overset{\displaystyle R_1}{|}}{C}H-CO-NH-\overset{\overset{\displaystyle R_2}{|}}{C}H-CO-NH-\overset{\overset{\displaystyle R_3}{|}}{C}H-\ldots$$

where the groups shown as R_1, R_2 and R_3 will be the different types of amino acid *side chains*. Looking at Table 2.3 we see that these can be hydrophobic or hydrophilic, positively or negatively charged or uncharged, and large or small. Different combinations taken from these therefore result in proteins with all sorts of different properties. When building up a polypeptide chain we have a choice of 20 amino acids for the first in the chain, of 20 for the second in the chain, and so on. The number of polypeptide chains it is possible to build is therefore $20 \times 20 \times 20 \times 20 \ldots$. A typical polypeptide chain may contain several hundred amino acids and therefore the number of possibilities is astronomical.

What shapes are protein molecules?

Although protein molecules are basically long, unbranched chains of amino acids, they may not necessarily occur naturally as long thin molecules. Most of the soluble proteins tend to coil up to form more or less globular structures, while many of the structural or insoluble proteins tend to twist together to form ropes or lie side by side to form sheets (see p. 84). Enzymes, protein hormones and antibodies are soluble proteins and we shall concentrate on these in the present section. However, we should remember that the different properties shown by the structural, insoluble proteins simply reflects their different amino acid side chains.

Boiling proteins with acid hydrolyzes the peptide bonds releasing the individual amino acids. Following such a hydrolysis it is possible to determine the amino acid composition of a protein. Nowadays this analysis is usually performed on an instrument called an 'amino acid analyzer'. Such instruments are highly automated, but their basic principle of operation is that of separating the different amino acids by column chromatography.

When we find out the amino acid composition of a protein, especially a soluble protein, by performing such an analysis, it tells us surprisingly little about the protein and its properties. This is because such an analysis tells us neither the *sequence* of the amino acids nor the *shape* of the protein molecule, and these are crucial to its biological function. There are good reasons for believing that the sequence of amino acids in proteins to a large extent controls shape. At present, however, we are not able to predict shape very accurately from a knowledge of sequence alone.

A major step forward in biochemistry occurred in 1953 when F. Sanger determined the sequence of the protein hormone, insulin (Figure 2.9). The techniques developed by Sanger have since been used on several hundred proteins and we now know the amino acid sequences of many proteins, some of them very much larger than insulin (Box 2.4). In parallel with the sequencing of proteins, it has been possible to discover the *shapes* of many proteins by using the technique of *X-ray crystallography*. These two analytical techniques complement one another: the amino acid sequence does not tell us much about the shape and the shape does not tell us much about the sequence.

Taken together, however, they enable us to build up a complete picture of the structure of soluble protein molecules.

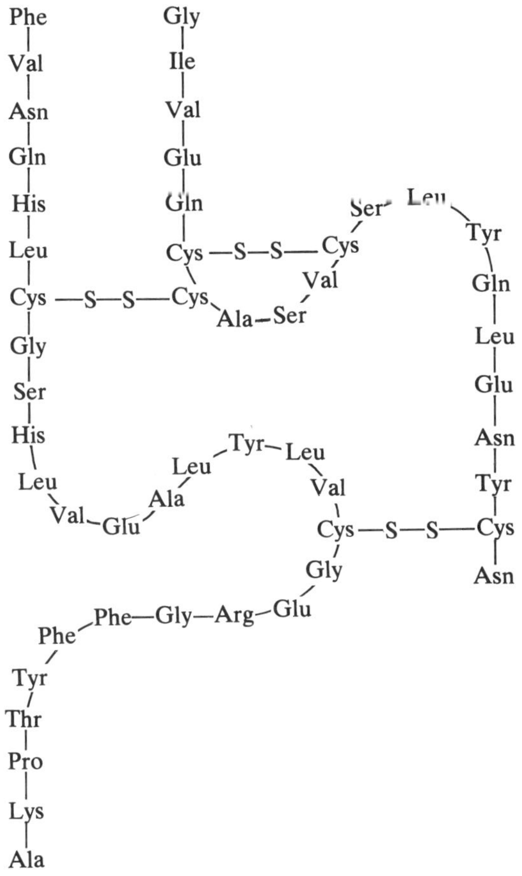

Figure 2.9 The amino acid sequence of ox insulin. Insulin is made up of two short polypeptide chains with 21 and 30 amino acid residues, respectively. The two chains are linked together by disulphide bridges

Box 2.4 *Determining the sequence of amino acids in a protein*

The technique for sequencing a protein depends upon a few specific chemical reactions and upon chromatography. The steps are: to label the end of the polypeptide chain by means of a chemical reagent, to hydrolyze the whole chain to its constituent amino acids, and then to identify the one amino acid with a chemical label attached by means of paper chromatography. The reagent fluorodinitrobenzene will react with the free amino (NH_2) group at the end of a polypeptide chain. The reagent is yellow in colour, and therefore after hydrolysis it is simply necessary to pick out, and identify, any amino acid which moves as a yellow spot on the chromatography paper. This technique, of course, only identifies the position of a single amino acid and is known as 'end-group analysis'.

It would be very convenient if chemical reactions were available whereby amino acids could be removed one-by-one from the end of a polypeptide chain and identified. This *can* be done to some extent and some modern instruments called 'sequenators' do this automatically. However, for various reasons, after 10–20 amino acid residues have been sequenced there begins to be great uncertainty about the identity of the 'next' residue. If we remember that a typical polypeptide is several *hundred* amino acids long, it is obvious that a strategy must be devised to overcome this problem.

The strategy usually adopted is to break up the polypeptide into a number of pieces by the use of chemical hydrolysis or proteolytic enzymes such as trypsin. The resulting peptides are separated by chromatography and electrophoresis. (This technique is called peptide mapping and is described in greater detail in Box 5.8.) The smaller peptides obtained are each sequenced and then the sequences are pieced together to get the sequence of the original polypeptide. It is usually necessary to use two or more cleaving reagents which split the chain in different places. In this way, overlapping peptide sequences can be identified showing how the small peptides are joined in the original polypeptide. The following very simple example shows how a sequence is built up:

1. End-group analysis identifies amino acid at NH_2 end as *tyrosine* (Tyr)
2. Cleavage with trypsin gives three peptides:

    Tyr—Lys   Glu—Met—Leu—Gly—Arg   Ala—Gly

3. Cleavage with a chemical reagent gives two peptides:

    Tyr—Lys—Glu—Met   Leu—Gly—Arg—Ala—Gly

4. Therefore sequence is:

    Tyr—Lys—Glu—Met—Leu—Gly—Arg—Ala—Gly

Levels of protein structure

In solution a typical soluble protein is folded into a more or less globular shape. Certain features of the folding can usually be distinguished, and the best known and most common of these is the α-helix. Because of the limited rotation around the bonds on either side of the peptide bond, only a few angles for these bonds are energetically stable. When one particular set of angles recurs the result is an α-helix (Figure 2.10): in this structure each complete turn of the helix takes up approximately 3.6 amino acid units. The structure is stabilized by weak bonds, in this case hydrogen bonds. Such structural features as this are referred to as *secondary structure*, distinguished from the amino acid sequence itself which is the *primary structure* of a protein. Another type of secondary structure found is the β-pleated sheet structure. This occurs more commonly in the insoluble, structural proteins, but regions of β-structure are found in soluble proteins too. It is characterized by hydrogen bonding between sections of polypeptide chain lying side by side (Figure 2.11). In addition, in a typical soluble protein, sections of α-helix and β-pleated sheet may be linked by sections of polypeptide chain in which no particular structural arrangement is obvious. These sections are probably just as important as the others in giving the protein its characteristic folded shape. The resulting three-dimensional folding of the polypeptide chain, super-imposed on α-helical and β-pleated sheet regions is called the *tertiary structure* of a protein. In forming this three-dimensional structure covalent cross-links

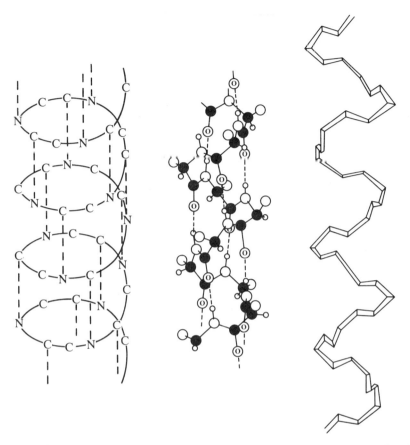

Figure 2.10 Schematic diagram to show the peptide backbone in the α-helix. Note how hydrogen bonds can form between adjacent turns of the helix. The *pitch* of the helix, the distance between two successive turns, is 5.4 Å, and there are 3.6 amino acid residues per turn. The 'hole' down the centre of the helix is too narrow to allow penetration by solvent molecules and this is probably a factor leading to the stability of the α-helix

between parts of the chain are often formed. These links are typically through two atoms of sulphur present in the amino acid cysteine. To some extent such links impose an overall shape to the molecule, but probably more important are the weak forces between different amino acid residues and peptide bond atoms. Typically hydrophobic amino acid side chains tend to come together on the 'inside' of the globular molecule, trying to keep away from the aqueous solvent. On the 'surface' of the molecule hydrophilic and charged residues will be found. These can form hydrogen bonds and ionic links, and play an important role in stabilizing the whole structure. Although this is a typical arrangement for globular, soluble proteins, almost exactly the opposite is found in proteins whose function depends on their 'sitting' in membranes (see p. 22). Here the hydrophobic amino acids are on the outside with hydrophilic residues on the inside!

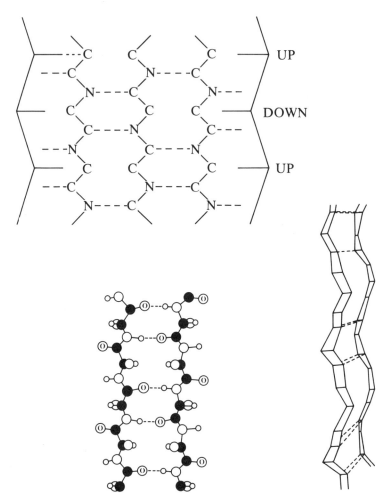

Figure 2.11 Schematic diagram to show the arrangement of polypeptide chains in the β-pleated sheet structure. This type of structure is found in the protein of silk called fibroin. However, within globular proteins, regions of β-pleated sheet structure may be found

Are 'polypeptide chain' and 'protein' synonymous?

Some proteins are composed of a single polypeptide chain but many have more than one polypeptide chain. A well-known example is the oxygen-carrying protein of the blood, haemoglobin. This is made of four polypeptides alike in pairs, two α and two β, which fit together quite tightly, but not covalently, to form a more or less globular molecule. The folded polypeptide chains are called 'subunits'. Proteins which are built up from subunits are said to have *quaternary structure*. The possession of such structure implies that the subunits fit closely to one another, i.e. they have complementary shapes, so that the whole can be stabilized by weak bonds.

Proteins having two or four subunit polypeptide chains are common, but others are known which have over twenty subunits, and virus coat proteins (see Box 1.5) may be considered to be multi-subunit proteins.

How stable are proteins?

The complex structure of a protein is maintained by the interplay of very many weak bonds. Because of the constant bombardment of a protein in solution by solvent molecules due to thermal energy, we must look upon a given structure as flexible or dynamic, not having an exactly fixed shape. It is likely that many of the properties of proteins depend on exploiting this small degree of flexibility.

Because protein structure is maintained by weak bonds, proteins tend only to be stable under rather mild conditions. We speak of proteins, as they occur and function naturally, as being in the *native state*. The mild conditions referred to are at temperatures between 0 and 40 °C, in the pH range from 5.5 to 9.0 and in the absence of organic solvents. Outside this range of conditions the weak forces holding the structure together become disrupted and this may result in a permanent change in the structure of the protein. This is referred to as *denaturation*. We observe this phenomenon whenever we boil an egg. The white of an egg is a concentrated solution of protein and the boiling process imparts so much thermal energy to the protein that its dynamic, soluble conformation is permanently lost and we get a white opalescent gel. This gel is actually made up of precipitated protein with entrapped water.

It is important to note that the process of denaturation does *not* involve splitting amino acids from the polypeptide chain. Breaking of the covalent peptide bonds requires much more 'violent' conditions such as boiling with 6 M HCl for 24 hours. However, denaturation *does* involve loss of biological activity. Thus boiling a solution of an enzyme will almost always result in loss of catalytic activity. Nevertheless, in many instances denaturation can be reversible. Some proteins can be precipitated from solution by the addition of ice-cold acetone or ethanol. If such precipitates are carefully dried to free them of the organic solvent, they can be redissolved in water with full restoration of biological activity.

This shows us that the primary structure largely controls the final three-dimensional arrangement of the polypeptide chain *and* that the three-dimensional structure is responsible for the biological activity of the native protein. Denaturation of proteins, by heat, strong acids or alkalis, organic solvents, etc., is an 'unphysiological' process, that is, one that rarely occurs in the normal functioning of a living organism. Indeed the internal environment of cells is tightly regulated so that the prevailing conditions are in fact 'mild', as defined above. The consequences of failing to maintain such conditions will be obvious if we accept that the life activities of organisms are totally dependent upon the biological activities of proteins such as enzymes, hormones, oxygen carriers, and so on. There are, however, examples of protein denaturation being truly physiological processes. When we eat protein-containing foods, the ingested protein upon reaching the stomach is immediately subjected to the gastric acidity, resulting in denaturation. This is no doubt important in making the protein more susceptible to digestion by the enzymes of the intestine. It is probably also important in *destroying* the biological activities of ingested proteins. Pepsin is a rather unusual protein, having been 'designed' to operate at very low pH values (see Figure 2.15).

Conjugated proteins

Proteins by themselves perform many tasks in living organisms. Sometimes, however, additional 'bits' are added to the molecules to give them other properties. These additional 'bits' may be metal ions, small non-protein, organic molecules, often derived from vitamins, or sometimes both. Such additions are called *prosthetic groups* because they 'help' the protein, giving it properties it would otherwise not possess. Haemoglobin is a good example of a protein with a prosthetic group. Each molecule of haemoglobin is made up of four subunits and each subunit possesses a single iron atom in a complex organic molecule called a porphyrin which is embedded in a cleft in the protein subunit (Figure 2.12). It is, in fact, the iron molecule that binds oxygen enabling haemoglobin to function as an oxygen carrier. Without its prosthetic group haemoglobin (or 'globin') is helpless!

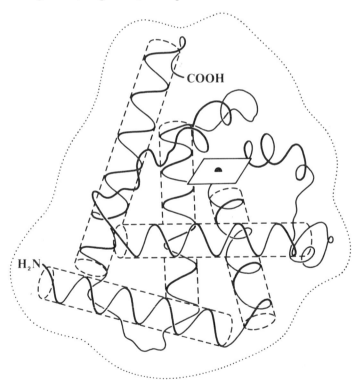

Figure 2.12 The haem prosthetic group lies in a hydrophobic cleft in the protein molecule of myoglobin and haemoglobin. The diagram, which is highly simplified, shows one subunit of a haemoglobin molecule. The dotted line is intended to indicate the overall shape of the folded polypeptide chain when the volume occupied by the amino acid side chains is taken into account. Note the extensive regions of α-helical structure

Functions of proteins in living organisms

Proteins play many roles in living organisms, from the purely structural to catalysis, carriage of oxygen, controlling by hormone action, etc. For the time being we will concentrate on what must be their most important function, that of catalysis, as enzymes. We should not let this obscure their many other functions, however. Most of the biological functions of the soluble proteins depend on their ability to *interact with* other molecules. Specific interaction means that molecules can be *recognized* and then appropriate action taken. In the case of enzymes, the sequence is: recognition and binding to the correct molecule, and then promotion of a chemical reaction, i.e. catalysis. Other examples of interaction will be found on p. 74.

2.5 Enzymes

The vast majority of chemical reactions within organisms only proceed at a measurable rate because of the presence of special catalysts called enzymes which are globular proteins. The fundamental importance of enzymes was recognized long ago by biochemists who saw that to understand the chemistry of living organisms it was necessary to understand how they performed all the chemical reactions essential for life.

Are all enzymes proteins?

The first enzyme to be purified to such a degree that it could be crystallized was *urease*, in 1926, by Sumner. Since pure urease appeared to be composed

$$CO\begin{array}{c}\diagup NH_2 \\ \diagdown NH_2\end{array} + H_2O \xrightarrow{\text{urease}} CO_2 + 2NH_3$$

urea   water   carbon   ammonia
                          dioxide

only of protein it was concluded that a protein *alone* could catalyze a chemical reaction. Since that time it has been amply confirmed that all the known enzymes (some several thousand of them) are proteins. They are all built up from amino acids, are susceptible to all the conditions that denature proteins described on p. 61 and are destroyed by digestion by yet other enzymes that hydrolyze proteins such as pepsin from the stomach. Although quite a number of enzymes consist solely of protein, a much greater number depend on help from prosthetic groups (see p. 69).

What is catalysis?

A catalyst is any substance which regulates the rate of a chemical reaction and itself remains unchanged. Catalysts cannot make reactions go that are energetically unfavourable: in fact catalysts merely speed the attainment of equilibrium.

Many reactions that are energetically favourable do not necessarily proceed rapidly because of the existence of the so-called *activation* energy. Petrol, in the presence of air, could, for example, undergo combustion with a massive and indeed explosive release of energy—obviously a very favourable reaction energetically. Yet it does not happen unless we add a spark. The spark raises the energy of a few molecules, effectively pushing apart some of the atoms in the petrol molecule which are covalently bonded together. This makes the bonds less stable and reaction with oxygen may then take place. When reaction has taken place much heat is released which can then raise the energy of yet more petrol molecules, and so on until an explosion occurs.

The barrier formed by the activation energy is very important to living organisms. If there were no such barrier, all the molecules of which we are made—carbohydrates, fats, proteins—and on which life depends, would promptly break down in the presence of oxygen in the air!

Although the activation energy barrier is important to living organisms, they must nevertheless have ways of overcoming this barrier whenever they *do* want reactions to take place. In the laboratory the standard way of getting chemical reactions to go at an appreciable rate is to supply heat. Normally, only a very few molecules have sufficient energy to react: adding heat increases the average energy of the molecules so that more have sufficient energy to react. Living organisms cannot apply heat: if they did they would destroy themselves by denaturing their proteins, disrupting their membranes, and so on. They must do 'cold chemistry' and this is why biological catalysts or enzymes are used. Instead of raising the average energy of the reacting molecules by supplying heat, the effect of a catalyst is to lower the activation energy barrier.

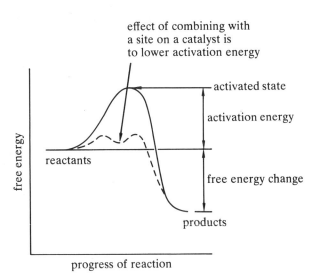

Figure 2.13 Energy changes during the course of an exergonic reaction. The activation energy forms a barrier to reaction, although the overall energy change for the reaction is favourable. Heat raises the average energy of the molecules so that more of them have sufficient energy to get over this barrier. Alternatively, formation of a complex on the surface of a catalyst brings the reactants into close proximity favouring reaction. The effect is an apparent lowering of the activation energy for the reaction. Note that the *total* energy change for the reaction remains unchanged

Chemists also use catalysts such as finely divided nickel, platinized asbestos and many others. These are thought to function by providing a *site* upon which substances can react. Heat, of course, speeds up reactions by increasing the thermal energy of the reactants so that more effective collisions take place per unit time. If, on the other hand, a *site* is provided to which the reactants are attracted and held in close proximity, then effective collisions become more likely. Although we do not completely understand how enzyme catalysis works there is no doubt that there are sites at which reactants bind to the protein and at which catalysis takes place. Thus the comparison with purely chemical catalysis is probably valid. There is also no doubt that the principle of reduction of activation energy by catalysis is fundamental to life (Figure 2.13). As a catalyst is unchanged at the end of a reaction very small amounts of catalyst, or enzyme protein, can be responsible for very large amounts of chemical transformation.

Enzyme specificity and the active site

The majority of the catalysts used in chemistry are relatively non-selective about which reactions they catalyze. By contrast, enzymes differ in two ways. Firstly, a given enzyme will typically deal with only one set of reactants (these are called *substrates* of the enzyme). In other words, enzyme catalysts have an extremely high degree of specificity compared with 'chemical' catalysts. Secondly, of the various possible reactions of the substrates, only one particular pathway proceeds under the influence of the enzyme catalyst. The result is to 'steer' substrates into a particular reaction pathway, with very few by-products. This is essential as living organisms do not have the ability to 'purify' the compound they want at each stage of a series of reactions as is typically done by an organic chemist.

Is enzyme catalysis a surface phenomenon?

If we compare the size of the average substrate molecule with that of the average enzyme molecule we reach the conclusion that the substrate can only be in contact with a small part of the enzyme molecule. Experiments show that typically an enzyme molecule only deals with one substrate molecule at a time. It is therefore believed that there is a certain region on the surface of the enzyme molecule which is called the *active site*. This is defined as the region to which the substrate(s) attach and upon which catalysis takes place. Most enzymes have one active site per protein molecule, but some have two or four sites, usually on separate polypeptide chain subunits. In a rather similar way a molecule of haemoglobin (although not an enzyme) has four 'sites' for binding oxygen.

Effect of temperature and pH

The conclusion that enzyme action depends on the three-dimensional shape of the enzyme protein molecule is consistent with the finding that when the three-dimensional structure is disorganized by denaturation, catalytic activity is lost. We see this when we observe the effect of temperature on the rate of an enzyme-catalyzed reaction (Figure 2.14). Similarly, enzyme activity is very

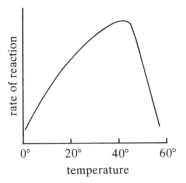

Figure 2.14 Effect of temperature on an enzyme-catalyzed reaction. With increasing temperature the average energy of the reacting molecules increases so that the rate of reaction increases. However, at around 55 °C (for most enzymes) the thermal energy causes the weak bonds that maintain the protein in an active configuration to start to break, and denaturation takes place with loss of catalytic activity. Above about 55 °C the rate of denaturation becomes more and more rapid

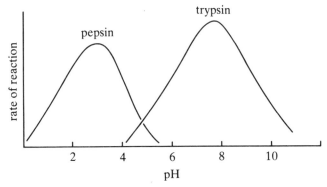

Figure 2.15 Effect of pH on the rate of an enzyme-catalyzed reaction. Most enzymes are at their most active (pH optimum) at about the pH value of the interior of living cells. Outside the normal pH range activity drops off because the ionization state of the amino acid side chains in the enzyme protein changes making catalysis less effective. At the extremes of pH (strong acid, strong alkali) the enzyme protein is denatured. A few enzymes are designed to operate at pH values far removed from intracellular ones. The best-known example is the proteolytic enzyme of the stomach, pepsin. In contrast, the proteolytic enzyme of the small intestine, trypsin, has a 'normal' pH dependence

sensitive to changes in pH. Enzymes are only active over a particular pH range, often with an optimum pH for activity (Figure 2.15). Outside this range activity is completely lost. Changes in pH result in breakage of the weak bonds that stabilize the three-dimensional structure of the native protein. It may also change the ionization state of amino acid side chains involved in the functioning of the active site.

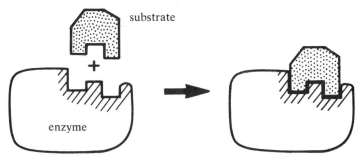

Figure 2.16 The 'lock and key' hypothesis. It was originally proposed that substrate molecules fit into the active site of an enzyme like a key fits in a lock

Structure of the active site and catalysis

The relationship between substrate and active site was seen traditionally as being like that between lock and key (Figure 2.16). At least part of the substrate molecule and the enzyme's active site must fit together in space intimately enough to become temporarily bonded forming a transient *enzyme–substrate complex*:

enzyme + substrate ⇌ enzyme–substrate complex → enzyme + products

This so-called 'lock and key hypothesis' is believed to be essentially correct, but it does not completely explain the observed behaviour of the enzyme during catalysis. For this reason the 'induced fit hypothesis' was suggested more recently (Figure 2.17). This theory proposes that both active site and substrate are to some extent flexible and that the structure of the site only approximates that of the substrate. However, when combination takes place to form the enzyme–substrate complex a small change occurs in the shape of the enzyme which 'improves' the fit and makes the enzyme–substrate complex highly reactive. From what we know of the detailed structure of a few enzymes it seems that the induced fit hypothesis explains the facts better than the lock and key model (Figure 2.18). However, we still have a great deal to learn about enzyme catalysis and enzyme specificity.

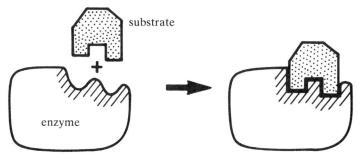

Figure 2.17 The 'induced fit' model. Not all the properties of enzymes could be explained by the lock and key hypothesis. It seemed more likely that when the substrate combines with the enzyme it induced a small change of shape of the enzyme protein molecule. Because both enzyme and, to some extent, substrate are distorted in the resulting complex, reaction takes place rapidly, yielding products

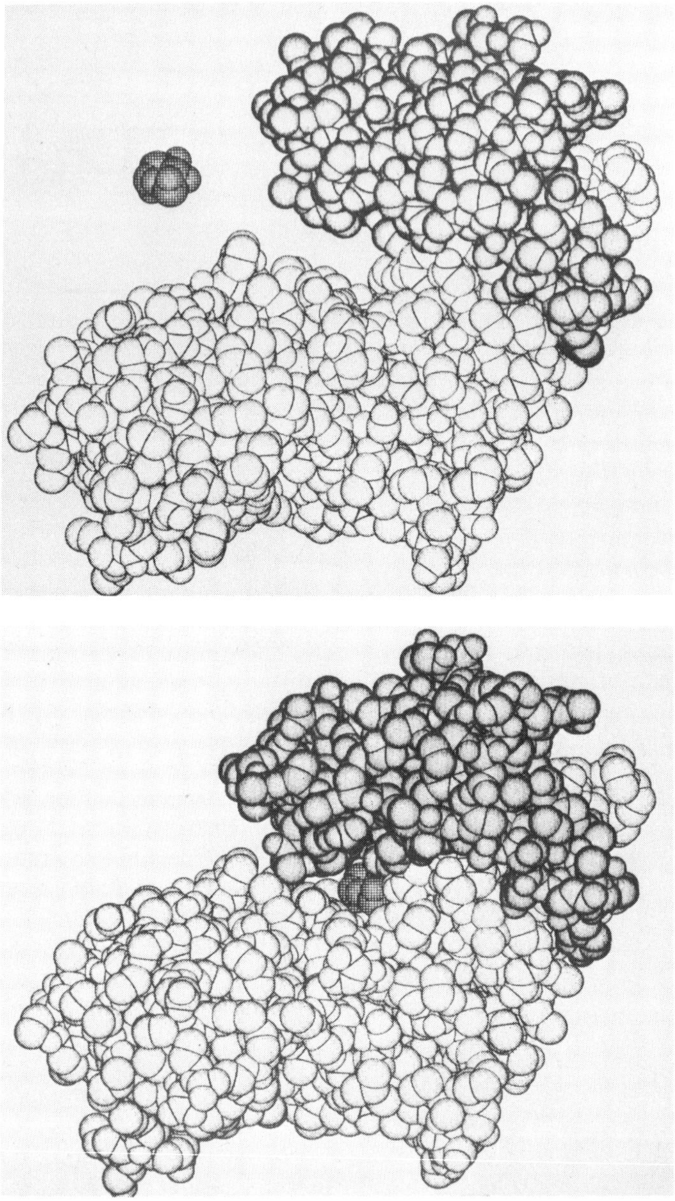

Figure 2.18 An enzyme combines with its substrate. The enzyme here is hexokinase from yeast which catalyzes the reaction:

$$glucose + ATP \rightarrow glucose\ 6\text{-}phosphate + ADP$$

The computer-produced space-filling models show hexokinase combining with a molecule of glucose, the first step of the reaction. The substrate (glucose) induces a change of conformation in the polypeptide chain. The small lobe of the protein molecule (more heavily shaded) rotates by 12° relative to the large lobe, closing the cleft between the lobes. (*Drawing kindly supplied by Professor T. A. Steitz, Yale University*)

Prosthetic groups, cofactors and coenzymes

It was mentioned above that some proteins have prosthetic groups to help them perform their tasks. Such groups are frequently found in enzyme proteins and have been called *coenzymes*, and sometimes *cofactors*. Since a very wide range of such molecules is known, functioning in many different ways, it is sometimes difficult to decide what should be called a coenzyme and what should be called a prosthetic group. No hard and fast definition can be given which applies to all situations, and as we proceed we shall be more interested in understanding the chemical mechanisms by which organisms operate than in trying to make rigid classifications. Many cofactors are derivatives of vitamins, essential dietary components (Box 3.9). It should be stressed, however, that there is no sharp line dividing relatively tightly-bound prosthetic groups from relatively loosely-bound coenzymes. Both provide specific chemical properties not attainable with amino acid residues alone. In addition, coenzymes can act as 'handles' enabling specific recognitition to take place (see p. 103).

When we look at the action of cofactors we find that not only do they dramatically increase the capacity or specificity for binding substrates, but also they may go through a cycle of reactions during the catalysis itself. A cofactor may, for example, accept protons and electrons from one substrate and then in a subsequent reaction pass these on to a second substrate, thus participating in an oxidation–reduction reaction. Cofactors may be categorized, albeit artificially, as:

(1) *Prosthetic groups*—those which complete their catalytic reaction cycles while attached to the same enzyme.
(2) *Coenzymes*—those which require the participation of two enzyme forms in the course of their catalytic cycle.

Enzyme classification

A long time ago the digestive enzymes of stomach and small intestine were given names ending in *-in*: thus peps*in*, tryps*in* and chymotryps*in*. On this basis certain other enzymes which had a protein-digesting action were also given names ending in the same suffix: thus fibrinolysin and cathepsin. However, the more widely accepted way of naming enzymes has been to add the suffix *-ase* to the reaction catalyzed. Thus a hydrol*ase* catalyzes a hydrolysis and a dehydrogen*ase* catalyzes the removal of hydrogen, and so on. The more recently discovered digestive enzymes were named according to this scheme. A peptidase is an enzyme that catalyzes the cleavage of peptide bonds. Thus carboxypeptid*ase* from the pancreas catalyzes the removal of amino acids from the carboxyl end of polypeptide chains.

A rigid numerical classification for all enzymes has been devised and is widely used in the scientific literature (Box 2.5). It recognizes that there are only six basic types of reaction catalyzed by enzymes and gives an Enzyme Commission Number to every enzyme. Although such a classification is important for specifying precisely an enzyme and its source, you will find that the typical biochemist going about his daily business in the laboratory speaks not of 'EC 3.4.4.5' but of 'chymotrypsin'.

Box 2.5 Enzyme nomenclature

The International Union of Biochemical Societies has recommended a numerical system of classification for enzymes. Enzyme activities are divided into six broad types or classifications and within these, reactions are further classified. Some examples are given below:

1. *OXIDOREDUCTASES*:	e.g. 1.1. Acting on CHOH groups of donors
	1.2. Acting on the aldehyde or keto groups of donors
	1.8. Acting on sulphur groups of donors
2. *TRANSFERASES*:	e.g. 2.6. Transferring nitrogenous groups
	2.8. Transferring sulphur-containing groups
3. *HYDROLASES*:	e.g. 3.1. Acting on ester bonds
	3.4. Acting on peptide bonds
4. *LYASES* (Bond-breaking reactions)	e.g. 4.3. Carbon–nitrogen lyases
5. *ISOMERASES*:	e.g. 5.2. *Cis-trans* isomerases
6. *LIGASES*: (Bond-forming reactions)	e.g. 6.4. Forming C—C bonds

Within each group further subdivisions are made so that each enzyme is classified by a set of four numbers thus:

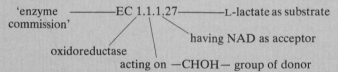

'enzyme ————EC 1.1.1.27————L-lactate as substrate
commission'
                                    having NAD as acceptor
    oxidoreductase
         acting on —CHOH— group of donor

Although this system is precise and unambiguous you will find that in every day usage, biochemists do not speak of EC 1.1.1.27 nor even 'L-lactate: NAD oxidoreductase', but rather call the enzyme 'lactate dehydrogenase'.

Measuring enzyme activity

Enzyme activity is measured by determining how much substrate is converted to product per unit time. Any appropriate chemical method may be used, depending upon the nature of the substances involved. Either rate of disappearance of substrate or rate of appearance of product may be followed. When making such a measurement it is found that the rate falls off with time unless fresh substrate is continually added. This is because the reaction is proceeding towards equilibrium and the effective substrate concentration is falling continuously from the moment the reaction is started (Figure 2.19). This shows us that the rate of enzyme reaction depends upon substrate concentration. It follows therefore that we should use initial rates when expressing enzyme activity. The units in which enzyme activity is expressed might be grams of substrate converted per minute or mol of substrate converted per second, per unit of mass of enzyme. If, with a fixed amount of enzyme, we do a series of experiments at different substrate concentrations we obtain a series of initial rates. Plotting initial rates against substrate concentration gives a graph of the type shown in Figure 2.20. It is seen that

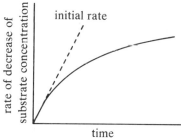

Figure 2.19 A progress curve for an enzyme reaction. The initial rate is shown. At any time after this the rate is lower because the concentration of substrate has been reduced by the action of the enzyme

above a certain substrate concentration the curve flattens out and no further increase in rate occurs. This simply means that the enzyme is working as fast as it can under those conditions. All the enzyme molecules are in the form of an enzyme–substrate complex and as soon as one reaction is finished another substrate molecule is ready to come in and form a new enzyme–substrate complex. Under these conditions we say that the enzyme is *saturated* with substrate. (It should be obvious that if, at a given substrate concentration, we increase the enzyme concentration, then the reaction will go faster: we are simply increasing the number of catalyst molecules available.)

The curve of rate of enzyme reaction against substrate concentration is a very important one. The curve is described as a *rectangular hyperbola* and two points on the graph serve to characterize it. One is the point at which the maximal rate of reaction is reached (see Figure 2.20), usually called V_{max}, the maximal velocity. The other point is the substrate concentration required to give half the maximal velocity. This point is referred to as K_m, the Michaelis Constant after Leonor Michaelis who, with Maud Menten in 1913, proposed a simple model to explain the kinetics of enzyme catalysis. The K_m is a constant for a particular enzyme. It measures the degree of attraction, or 'affinity', of enzyme for substrate: the smaller the K_m the higher the affinity and *vice versa*. We shall see shortly how the measurement of K_m is of value.

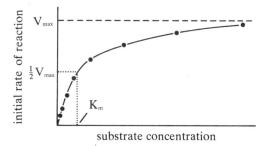

Figure 2.20 Plot of initial rate of an enzyme-catalyzed reaction against substrate concentration. At very high substrate concentrations the graph flattens out because catalysis is taking place at the maximal velocity (V_{max}). Adding further amounts of substrate does not increase the rate therefore. In practice it is very difficult to reach V_{max}, and points in this region of the curve tend to be inaccurate because of the difficulty of measuring initial rates accurately. The Michaelis constant, K_m, is the substrate concentration at half-maximal velocity

Inhibition of enzyme activity

A very important part of the science of 'enzymology', the study of enzymes, is the investigation of reagents that inhibit enzyme activity. Such investigation is not purely academic. A knowledge of inhibitors may tell us what reactive groups are present in the active site and may tell us something about the shape of the active site. This is because a substance with a similar but not identical shape to the true substrate may act as an inhibitor: a survey of the shapes of inhibitors may therefore allow us to decipher the shape of the site. Similarly, reagents which react with amino acid residues in the active site will inhibit the enzyme allowing us to determine which types of residue are present in the site. There are, however, other reasons for studying the enzyme inhibition. Many drugs, pesticides and herbicides exert their action by acting as enzyme inhibitors and there is thus much commercial interest in finding new inhibitors for use medically and agriculturally. In addition, the fact that there are so many different types of enzymes in even the simplest cell, means that mechanisms have evolved for controlling enzyme activity in cells. Such mechanisms also involve inhibition of enzyme activity.

Several types of inhibition are known, but we will concentrate on just two examples. In a way these illustrate the extremes of a range of types of inhibition. A compound with a similar structure to the substrate of an enzyme may be able to occupy the active site preventing catalysis taking place. Such inhibition is referred to as *competitive* because the inhibiting molecule can be

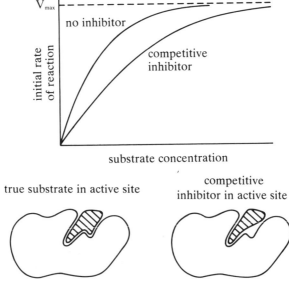

Figure 2.21 Inhibition of an enzyme by a competitive inhibitor. The shape of the inhibitor molecule is similar to that of the true substrate, hence it can get into the site and block catalysis. However, this is reversible, and when very many more substrate than inhibitor molecules are present, it is more likely that a substrate rather than an inhibitor molecule will find its way to the active site. At high substrate concentrations therefore inhibition is overcome and V_{max} is eventually reached even in the presence of inhibitor

'competed off' the site. If the substrate concentration is increased to a sufficiently high level, there is a greater chance of a substrate molecule getting to the active site than an inhibitor molecule. Therefore, at high concentrations of substrate it *is* possible to reach V_{max} (see Figure 2.21) *even in the presence of the inhibitor.* The effect is as though K_m increases. We would therefore say that the affinity of enzyme for substrate is apparently reduced in the presence of competitive inhibitor.

The second category of inhibition is *non-competitive inhibition.* In this situation it is thought that the inhibitor molecule binds at or near the active site blocking access of the substrate. The non-competitive inhibitor need be nothing like the true substrate in shape, but simply needs to link to some regions of the protein near the active site so blocking it. It should be stressed that in both cases the linking of inhibitor to enzyme molecule may be non-covalent and is reversible. In the case of competitive inhibition only, the inhibition is reversed at high substrate concentration simply because both substrate and inhibitor compete for the same site.

A non-competitive inhibitor (Figure 2.22) has no effect on the K_m, i.e. it does not change the affinity of the substrate for the enzyme. However, it does effectively put out of action a certain fraction of the molecules of enzyme, thus V_{max} falls: those enzyme molecules that are left behave 'normally'.

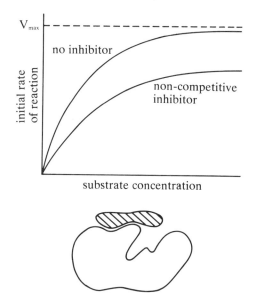

Figure 2.22 Inhibition of an enzyme by a non-competitive inhibitor. The inhibitor is not shaped like the true substrate and merely combines with some region in or near the active site preventing access by the substrate. There is no competition. The effect is as if the enzyme were reduced in concentration because a certain proportion of the enzyme molecules are inactive at any given time. V_{max} is never reached. (Compare with Fig. 2.21)

Control by inhibition

A special example of enzyme inhibition is *allosteric inhibition.* An allosteric

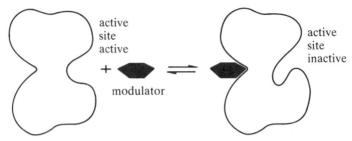

Figure 2.23 Allosteric inhibitors (modulators) cause a conformational change in the enzyme protein that distorts the active site leading to loss of activity. Other types of modulator can have positive effects, i.e. the active site only displays catalytic activity when the modulator is bound to the allosteric site

enzyme is one that can exist in two distinct spatial conformations, one catalytically active, the other inactive. There is a second site on the molecule which does not have catalytic activity, but merely has the ability to bind another molecule. The act of binding at this second site causes a conformational change to occur from active to inactive forms of the enzyme or *vice versa* (Figure 2.23). It is perhaps better to call the compound that binds at the second site a 'modulator' rather than an inhibitor since its effect can be either positive (i.e. turning on activity) or negative (i.e. turning off activity). Although these ideas are complex you should be able to see how such mechanisms are potentially capable of regulating the activities of enzyme catalysts in cells and hence regulating metabolism.

Other interactions

Many other properties of soluble proteins depend upon their ability to interact with other molecules. Enzymes are a special case where interaction of enzyme active site and substrate, representing 'recognition' of an appropriate substance, is followed by catalysis of a chemical reaction. In other instances, the processes following recognition are other than catalysis. Antibody proteins, for example, are designed to recognize invading organisms (bacteria, viruses) thus 'pointing them out' to the defence systems of the body such as the white blood cells, that can inactivate them. Another system is the oxygen-carrying system of the body. This has a number of interesting and important features and so we consider it briefly here.

2.6 Haemoglobin and oxygen transport

The protein haemoglobin occurs in the red blood corpuscles of vertebrate animals, and it is one of the most extensively studied of all proteins. It is capable of binding oxygen reversibly. The molecule is made up of four subunits, each with an iron atom in its prosthetic group. Each subunit can bind one molecule of oxygen and therefore one molecule of haemoglobin binds four molecules of oxygen.

The process of oxygen transport involves binding oxygen at one location, the lungs, transporting it to the tissues, and releasing it (Figure 2.24). The

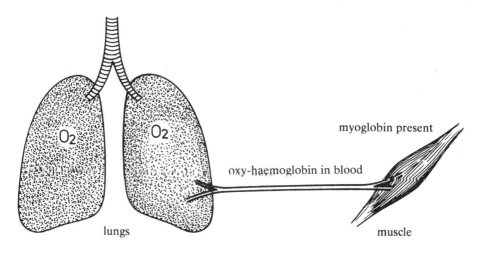

myoglobin present

oxy-haemoglobin in blood

lungs   muscle

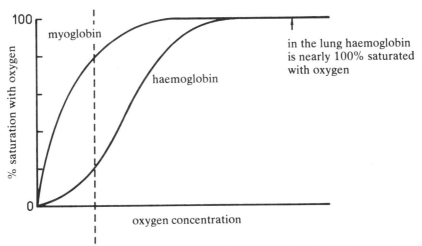

in the lung haemoglobin
is nearly 100% saturated
with oxygen

At this oxygen concentration (somewhere in the tissues of the body) haemoglobin would be 20% saturated but myoglobin would be 80% saturated. Therefore oxygen 'flows' from haemoglobin to myoglobin

Figure 2.24 Oxygen transport by haemoglobin

first requirement is that the binding should be specific, that is oxygen should be transported and not, say, nitrogen. The binding is specific to this extent, but it should be noted that other gases, such as carbon monoxide can be bound. When this happens the effect is similar to inhibiting an enzyme, especially as haemoglobin has a higher affinity for carbon monoxide than for oxygen. This, of course, prevents oxygen being transported and can result in death, for example, from a motor car exhaust in a confined space.

The contrast between the binding of oxygen and carbon monoxide illustrates an important point. The binding of the molecule to be transported, here oxygen, must be fairly loose so that oxygen can be released in the tissues at the appropriate time. This is indeed such an important feature that the molecule of haemoglobin incorporates mechanisms by which the strength of binding of oxygen is regulated and can be changed to meet the demands upon it. From studies of the haemoglobin molecule by X-ray crystallography and other methods, we know that the oxygen molecules are carried by haemoglobin because they combine loosely with the iron atom which is situated in a hydrophobic cleft in the protein. We know that when the first molecule of oxygen binds to one of the subunits of haemoglobin, it causes the iron atom with which it becomes closely associated to change its position very slightly. This very slight movement sends a little ripple of change through the polypeptide chain which is transmitted to the neighbouring subunits in the molecule. The result of this is that when the next oxygen molecule comes along, it is actually bound slightly more strongly. This process continues until four oxygen molecules are bound. Thus the affinity of haemoglobin for oxygen is changed step by step. We can see this if we compare the binding curves for haemoglobin and the monomeric but otherwise very similar oxygen-binding protein of muscle, myoglobin (Figure 2.24). We note that the binding curve for haemoglobin is S-shaped or sigmoidal while that for myoglobin is hyperbolic. If we measure the P_{50}, the equivalent of the K_m for these two proteins, that is the oxygen concentration at which the proteins are half-saturated with oxygen, we see that myoglobin has the highest affinity (i.e. lowest P_{50}).

The sort of change observed with haemoglobin when oxygen binds is another example of an allosteric control mechanism. Here the binding of a molecule at one site causes a conformational change that makes the remaining sites more reactive. This is referred to as co-operative binding. It is obviously not possible with a protein such as myoglobin which has only a single binding site for oxygen, and myoglobin always has a higher affinity for oxygen than haemoglobin. Therefore when blood is passing through the tissues, especially muscle, oxygen passes from haemoglobin to myoglobin. Myoglobin operates as an oxygen *store* deep in the muscle bed and it is noticeable that the muscle of diving animals such as seals and whales is deep red in colour, being very rich in myoglobin.

There are several ways in which changes in the oxygen affinity of haemoglobin, brought about by small molecules binding at sites other than the haem prosthetic groups, help the molecule function more efficiently. For example, addition of protons decreases the oxygen affinity, and this is known as the *Bohr effect*. Metabolically-active tissues tend to become slightly acid, as we shall see later. Therefore when blood passes through such tissues, the lowered oxygen affinity caused by the increased proton concentration frees more oxygen from the haemoglobin. The result is that oxygen delivery is increased selectively to sites with the greatest need.

In a somewhat different way, delivery of oxygen to the foetus is aided in pregnancy. Maternal and foetal blood pass close to one another in the placenta, and because the foetus is not using its lungs, its haemoglobin tends to be comparatively depleted in oxygen. Oxygen therefore tends to pass from maternal to foetal blood. This process is aided by the fact that the foetus produces its own type of haemoglobin which has a higher affinity for oxygen than normal, or adult, haemoglobin. Foetal haemoglobin levels in the blood start to decrease as soon as birth has taken place and is gradually replaced by 'adult' haemoglobin over a few weeks.

2.7 Structure

Finally in this chapter we consider the role of macromolecules in maintaining the structure of cells and organisms. All organisms have mechanical elements that give support, hold in cell contents or that function as tensile structures. Examples are bones, tendons, plant cell walls, insect exoskeletons and bacterial cell walls. In addition, there are materials that act as lubricants and that have water-repelling properties. These structural elements are built up from very diverse building blocks and so we ask if there are any ground rules.

What governs the use of a particular material as a structural element?

Two factors are involved. The most important of these is to obtain a structural material with the desired mechanical properties. The second is an economic consideration: the materials used should be comparatively easy to obtain and must be economical in raw materials. Thus plants, which are habitually short of nitrogen, rarely use nitrogen-containing structural materials. Plant cell walls indeed tend to be of cellulose or lignin, materials which contain practically no nitrogen. Relatively simple repeating polymers as we shall see suffice for the construction of these mechanical elements.

What factors govern the structures of supporting elements?

Long chain macromolecules are the basic structural materials in living organisms, and usually are polysaccharides or proteins. Sometimes they are used as such in tensile structures like tendons which are made of the protein collagen. By laying down fibres in different orientations, however, sheets and nets may be formed, and cross-links added later may strengthen such structures and give rigidity. Finally, a composite material in which strong fibres are buried in a bulk material has greater strength than either component by itself possesses. Such materials have shear (twisting) strength and rigidity coupled with a degree of flexibility. The secret is to be flexible enough to give way to small stresses and to prevent cracks from propagating. Fibreglass is such a composite man-made material with strong fibres embedded in an amorphous matrix. We shall see in what follows that various substances are employed in different situations to give different properties.

Plant cell walls and wood

The most common fibrous element in plants is cellulose which is probably the most abundant bulk organic chemical in the world. It is said that half the carbon in the world in organic combination is in the form of cellulose. Cellulose is formed from glucose residues linked $\beta 1 \rightarrow 4$ instead of $\alpha 1 \rightarrow 4$ as in starch or glycogen, and there is no branching (Figure 2.25). This apparently small difference, in practice, gives a great difference in physical properties. The arrangement allows much interchain bonding and water molecules cannot

(a)

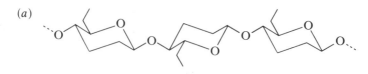

(b)

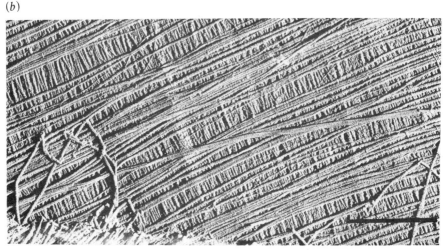

Figure 2.25 Structure of cellulose. Cellulose is a $\beta 1 \rightarrow 4$ linked polymer of glucose with no branching. (a) Shows the skeletal structure of the molecule. (b) Shows fibres of cellulose laid down at different angles in the cell wall of *Cladophora*. Length of line 1 μm. (*Electron micrograph kindly supplied by Professor R. D. Preston*)

penetrate. Cotton wool is practically pure cellulose and is completely insoluble in water. Cellulose, unlike starch, does not form a gel upon boiling with water.

For all its abundance, surprisingly little is known about the detailed structure of cellulose although it is known that the structure is almost crystalline. In plant cell walls, cellulose fibres are laid down in sheets with layers in which the fibres run in different directions. The network is cemented together by plant gums.

It is interesting that higher animals seem never to have evolved cellulases, enzymes capable of hydrolyzing the $\beta 1 \rightarrow 4$ links in cellulose, hence the valuable glucose of which it is made is unavailable to the vast majority of higher animals. Many micro-organisms produce cellulases and ruminant animals live off the waste products from cellulase-producing micro-organisms living in their stomachs. It is important to note that the micro-organisms do not release glucose for the animal's use, but break it down for their own purposes, finally releasing smaller molecules such as acetate and propionate the ruminant may use.

In mature wood, or xylem, a matrix material called *lignin* forms around the cellulose fibres. This is neither a polysaccharide nor a protein, although it is a polymer, and it contains practically no nitrogen. Lignin is derived from certain phenolic products which can polymerize with many cross-links (Figure 2.26). The resulting three-dimensional lattice with embedded cellulose fibres is extremely strong and durable. It is resistant to attack by all but a few fungi and insects, and the determination of its chemical structure is a daunting task.

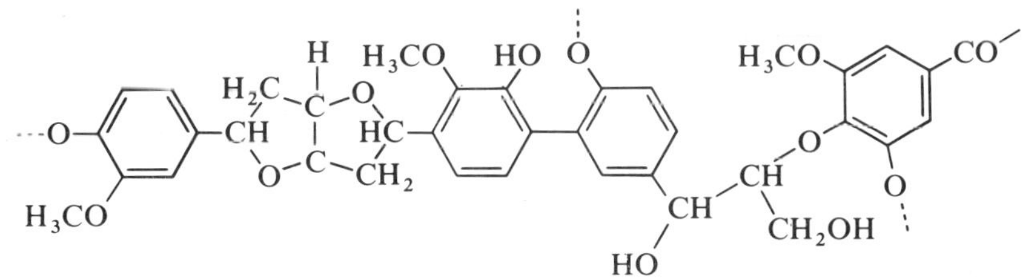

Figure 2.26 Structure of lignin. The diagram shows a very small section of the possible structure of spruce lignin. Note the large number of possible cross-links. The precursor for lignin is tyrosine from which the amino group is removed before polymerization

Invertebrate exoskeletons

Crustaceans and arthropods have *chitin* as the fibrous material in their exoskeletons. This is a nitrogen-containing polysaccharide made up of chains of N-acetyl-glucosamine residues (Box 2.6). Insect wing, which is the envy of engineers in terms of its strength and lightness, is almost pure chitin. For the tougher parts of the exoskeleton, however, the chitin fibres are embedded in a matrix of protein called *sclerotonin*. This is a tough, water-impermeable material, made so by *tanning*, that is, the protein chains are cross-linked by means of quinones to a greater or lesser extent to form a three-dimensional lattice. The quinones are formed from the amino acid tyrosine. The variations that can be played on this basic theme are enormous and a great range of different forms are found in arthropods. The amounts of chitin and sclerotonin may be varied, as well as the degree of cross-linking. Beetle cuticle, for example, tends to be highly tanned, and in crustaceans the inner layer of exoskeleton has a deposit of calcium carbonate crystals amongst the fibres making it similar in properties to vertebrate bone (see later).

Box 2.6 Chitin

Chitin is a polysaccharide formed by linking N-acetyl-glucosamine residues through $\beta 1 \rightarrow 4$ links:

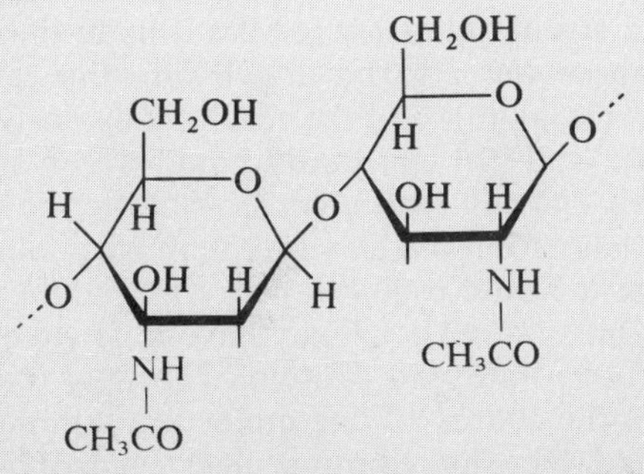

The cuticle of arthropods is formed by embedding the chitin fibres in a matrix of tanned protein called *sclerotonin*. The tanning process involves the generation of cross-links between protein molecules, and the tanning agents are quinones formed by oxidation of the amino acid tyrosine.

In crustaceans, the outer layers of the shell are tanned, and the inner layer has calcium carbonate deposited among the chitin fibres. This is comparable to the way in which bone is formed in vertebrates.

The compound eyes of insects have lenses of chitin. The electron micrograph below shows a section through one facet of the eye of the fruit-fly, *Drosophila melanogaster* (electron micrograph by courtesy of Dr. J. Sparrow).

Bacterial cell walls

Bacterial cells contain highly-concentrated solutions which tend to exert a very high osmotic pressure—some hundreds of atmospheres in some cells. Therefore, there is a strong tendency for the cells to burst because the high osmotic pressure causes water to be drawn into the cell. There are three possible ways of preventing this and the resulting explosive dilution: (a) have a water-impermeable wall, (b) constantly pump water out, or (c) withstand the pressure by fitting the cell with a corset. Solution (a) would hamper the taking up of materials from the environment and solution (b) would be uneconomical in that it would require energy to 'pump' water out. Bacteria have largely settled on solution (c). They have a net-like porous coat which is a polymer of carbohydrate and protein. The final cross-linking of these elements is such that the coat is all one molecule, with no weak points. Such coats have been called 'bag-shaped molecules' (Figure 2.27). There is plenty of room between the cross-links for molecules to get in and out of the cell. In addition, other materials may be present in the wall such as lipids and polysaccharides depending on the species of bacterium.

Certain bactericidal compounds exert their action on bacterial cell walls (Box 2.7). This illustrates the importance of the wall: destroy the wall and you destroy the bacterium.

Box 2.7 Penicillin, lysozyme and bacterial cell walls

The bacterial cell wall holds in the bacterial cell contents purely mechanically, acting as a net. This allows the cell to exist with a very high internal osmotic pressure (sometimes several hundred atmospheres). Removal of the cell wall under hypotonic conditions will lead to the cell bursting, and dying. Mechanisms which cause bursting are used to protect higher organisms against invading and potentially lethal bacteria.

Lysozyme is an enzyme found abundantly in egg white but also in tears, saliva and other secretions. Egg white lysozyme was the first enzyme for which the complete three-dimensional structure was determined by X-ray crystallography. The enzyme is a glycosidase: it cleaves the polysaccharide chains in the cell walls of certain bacteria. It is believed that when the substrate polysaccharide binds to the enzyme, distortion occurs, weakening certain bonds, so that hydrolysis takes place. Lysozyme is obviously part of the body's front line defence system being present in secretions to lyse bacteria *before* they can invade.

Penicillin was the first antibiotic to find practical use in medicine. It is a small organic compound produced by the fungus *Pencillium notatum* (see cover).

Several different penicillins are known which differ in the group *R*. Penicillin is most effective against gram-positive bacteria, but only kills actively-growing organisms. It works by preventing proper cell wall synthesis: the growing cells try to divide but then burst because they have an incomplete wall. The cell wall of certain bacteria contain the amino acid D-alanine (the alanine normally occurring in proteins is the L-form). It seems likely that the enzymes responsible for cell wall synthesis pick up penicillin by mistake for D-alanine—the structures are similar. Penicillin may work like a competitive inhibitor.

Penicillin is a very good antibiotic because it is almost completely non-toxic to humans. We have nothing resembling the bacterial cell wall structure nor the enzyme to make such a structure, and so penicillin, although highly toxic to growing bacteria, is almost completely non-toxic to us. This illustrates one of the basic principles of *chemotherapy*: for a compound to be suitable medically it should be highly toxic to the invading organism but harmless to the host animal.

Vertebrate structural materials—cartilage, bone and elastin

The connective tissue of the vertebrate body is built up from fibres of the protein, *collagen*, embedded in a polysaccharide matrix to form cartilage. The rods of collagen give the material flexibility and stiffness while the matrix holds them together. In *bone* much of the matrix is calcified, that is, it is replaced by crystals of calcium hydroxyphosphate, $Ca_5OH(PO_4)_3$, which imparts hardness and rigidity, and gives a load-bearing structure capable of supporting, for example, an elephant! *Elastin* is a rubbery protein in elastic fibres found particularly in the walls of blood vessels.

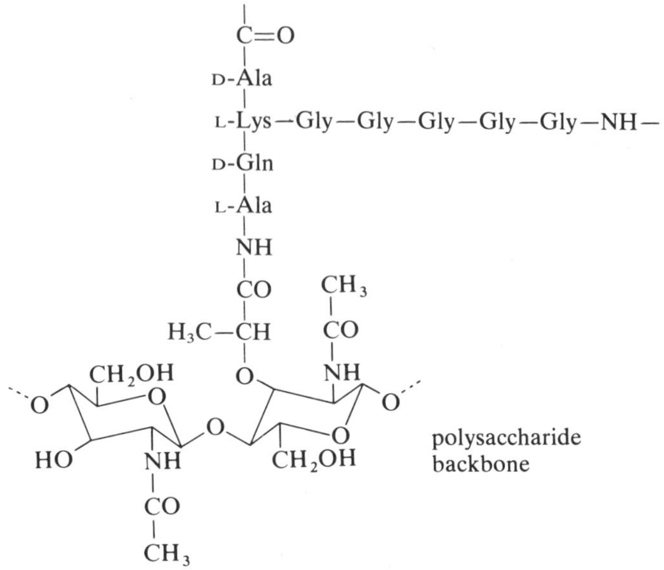

$$\begin{array}{c}
\mid \\
C{=}O \\
\mid \\
\text{D-Ala} \\
\mid \\
\text{L-Lys}-\text{Gly}-\text{Gly}-\text{Gly}-\text{Gly}-\text{Gly}-\text{NH}- \\
\mid \\
\text{D-Gln} \\
\mid \\
\text{L-Ala} \\
\mid \\
\text{NH} \\
\mid \\
\text{CO} \quad\quad \text{CH}_3 \\
\end{array}$$

H₃C—CH   CO

polysaccharide backbone

Figure 2.27 Structure of a bacterial cell wall. The wall is built up from a polysaccharide backbone, a tetrapeptide (L-alanine, D-glutamine, L-lysine, and D-alanine), and a penta-glycine bridge. The schematic diagram above shows how these form a continuous net. The bacterium in this instance is *Staphylococcus aureus*. Note the presence of abnormal D-amino acids which are never found in proteins. Other bacterial types have different amino acids and different carbohydrate units

Collagen

Collagen is the most abundant protein in the vertebrate body. It is insoluble and occurs in cartilage, bone, and also as a mat in the deep layers of the skin giving flexibility and resistance to tearing. It also forms the cornea of the eye. The different properties of these various structural tissues arise largely from the way in which the collagen fibres are laid down.

Collagen is a protein, but its composition is very different from that of the soluble proteins discussed previously. It shows, for example, a repeating structure. In the interior part of the polypeptide chain, every third residue is a glycine, and approximately one position in every five is occupied by the amino acid proline:

$$\ldots X-gly-pro-X-gly-X-pro-gly \ldots$$

The positions, X, are occupied by a variety of amino acid residues whose nature depends on the particular type of side chains. A typical composition might be: glycine 33%, proline and hydroxyproline (see below) 21% and alanine 11%. Mammalian collagen has a triple-stranded structure in which two out of the three strands are identical. Because of the presence of the many glycine residues (which effectively have no R side chain) and of the proline residues (which prevent the formation of α-helix), maximum bonding is developed by the formation of a triple helix in which the chains are twisted one turn for every three residues (Box 2.8). The resulting structure has a rope-like configuration.

When the collagen polypeptide chains are being synthesized in the cells, certain enzymes are present which convert about half of the proline residues to *hydroxyproline*. This sort of modification of a polypeptide chain after it has been synthesized is not at all typical of proteins and only occurs with collagen and a few similar structural proteins. It is not completely understood why this happens in collagen. It is also known that some of the lysine residues become modified to form hydroxylysine and that sugar residues can be added to these. Collagen is thus a *glycoprotein*: presumably these hydrophilic carbohydrate residues increase the interaction with the polysaccharide matrix in which the collagen fibres are buried.

Box 2.8 Collagen

Collagen is the most abundant protein in mammals and may constitute up to a quarter of the total protein of the body. It is the major structural protein of tendon, bone, cartilage and teeth, and also forms a strong network under the skin. Its insoluble fibres have a high tensile strength. However, collagen from the tissues of young animals is soluble, the reason being that the polypeptide chains have not yet become *cross-linked*. The electron micrograph below shows collagen fibres in a section of immature guinea-pig mammary gland (length of line 1 μm).

There are several types of collagen to serve different structural needs, but all are constructed as units in which three polypeptide chains twist together to form a *triple helix*. In type I collagen, the most common type, each triple helix is made up of two polypeptides of one type ($\alpha 1$) and one of another ($\alpha 2$). Other types of collagen have three identical chains. Each of the strands has about a thousand amino acid residues.

The amino acid sequence of the collagen polypeptides is very unusual compared with that of globular proteins. One-third of the residues are glycines and there are also many proline residues. In addition, two amino acids not found in other proteins are present, hydroxyproline and hydroxylysine, which are formed *after* the polypeptide chain has been synthesized. The sequence contains repeats, the sequence —glycine—proline—hydroxyproline— occurring frequently. In contrast, globular proteins rarely have repeating sequences; haemoglobin contains only about 5% glycine.

collagen triple helix

Amino acid sequence of part of a collagen polypeptide:

—gly—pro—ser—gly—pro—arg—gly—leu—hyp—gly—pro—hyp—gly—ala—

Sugars and cross-links. Carbohydrate units are added to hydroxylysyl residues, frequently as a disaccharide, glucose—galactose. These sugar residues are added to the immature form of collagen and the amount added depends on the type of tissue. The cross-links that form in collagen are covalent ones. The links are formed between lysyl side chains, and links are found both *within* a triple helix and *between* triple helices. The extent of cross-linking varies with the function and the age of the tissue. In a mature rat, the Achilles tendon has very many cross-links, whereas the tail tendon has comparatively fewer and is more flexible.

The polysaccharide matrix

The polysaccharide material in cartilage is very complex in structure: nevertheless a brief consideration of its chemical nature will help us to understand how it comes to have its particular properties. It is not a simple polysaccharide, but rather a complicated *protein–polysaccharide* called *chondroitin sulphate*. Collagen is a glycoprotein. This means that it is largely protein with a few carbohydrate residues attached covalently to it. Protein–polysaccharides, on the other hand, are formed from a small protein core with an enormous amount of carbohydrate attached. Thus they show the properties of polysaccharides more than those of proteins. In the formation of the matrix of cartilage, we start with a protein 'core' which contains an abundance of serine residues and on to this are built long carbohydrate chains to give a 'bottle-brush' type of structure. The sugar residues in this are derived from glucose and galactose. The glucose residues are modified to form *uronic acids* in which carbon 6 ($-CH_2OH$) has been oxidized to a carboxylic acid group ($-CO_2H$). The galactose residues are in the form of the sulphates of N-acetyl-galac-tosamine. This complicated structure has two important properties. Firstly, because of the negative charges on the carboxylic acid groups in the glucuronic acid residues ($-CO_2^-$) and the sulphate groups in the N-acetyl-galactosamine residues ($-SO_4^-$), the carbohydrate chains attached to the polypeptide core tend to repel one another. Secondly, because the carbohydrate chains are very hydrophilic, they attract and hold very many water molecules. The resulting molecule thus occupies a very large volume and is resilient. When mechanical pressure is applied, water molecules may be squeezed out, but will quickly find their way back in when the pressure is released. This strong but flexible type of structure changes to a strong, rigid structure when some of the matrix is replaced by calcium salts when bone is formed.

Questions

1. Write an essay on 'the structure and function of plant polysaccharides'.

2. The diagram shows the result of a chromatography experiment.

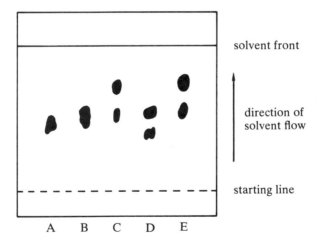

solvent front

direction of solvent flow

starting line

A   B   C   D   E

The following samples were spotted on the starting line:
A. The products from hydrolyzing starch with acid.
B. The products from hydrolyzing glycogen with acid.
C. The products from hydrolyzing sucrose with acid.
D. The products from hydrolyzing lactose (milk sugar) with acid.
E. A dilute solution of honey.
Answer the following questions:
(a) What is the approximate R_f value for glucose?
(b) Does lactose contain fructose? If not, what experiment could be done to determine its composition?
(c) What are the major sugars (monosaccharides) in honey?

86   Introducing biochemistry

3. This question is about biological techniques.
    (a) Describe how you would use chromatography to separate and identify *either* sugars *or* amino acids in a fruit juice.
    (b) Outline a method used to isolate mitochondria from liver tissue.

4. Which of the following statements is NOT true of both cellulose and proteins?
    A. Both are found in the cell membrane.
    B. Both can be used as energy sources by certain organisms.
    C. Both are compounds of high molecular mass.
    D. Both are synthesized by the condensation of a number of simple units.
    E. Both contain nitrogen.
    Give reasons for your choice.

5. What do you understand by the term *lipid*. Discuss the importance of lipids or lipid derivatives in cell membranes, prevention of desiccation and maintenance of body temperature.

6. The illustration shows the result of an experiment in which sickle cell haemoglobin was compared with normal human haemoglobin. The two haemoglobins were digested separately with the proteolytic enzyme trypsin. The resulting peptides were separated on sheets of paper by first electrophoresis and then, at 90°, by chromatography, to form a 'peptide map'. Finally, the peptides were made visible by staining.

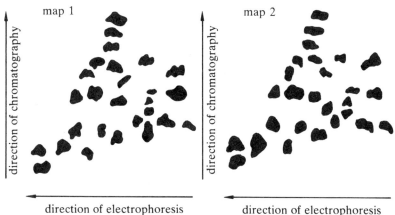

    (a) Sickle cell haemoglobin differs from normal haemoglobin by having a *neutral* valine residue replaced by an acidic glutamine acid residue. Explain how this accounts for the differences observed in the peptide maps.
    (b) Circle the peptide on map 2 (sickle cell) which has glutamic acid rather than valine.
    (c) Explain why trypsin treatment results in the production of discrete peptides rather than of random fragments or amino acids.
    (d) Mark with arrows any two peptides that separated during electrophoresis but which were not separated by chromatography.

7. Review the properties of enzymes including in your answer discussion of TWO of the following: (a) active sites, (b) inhibitors, (c) cofactors.

[O & C]

8. What are enzymes and how would you classify them? Give an account of their properties and discuss their mode of action.

[O & C] (S)

9. What is an enzyme?
Explain the mode of action of enzymes.
How could you measure the effect of EITHER substrate concentration OR enzyme concentration on the rate of a NAMED enzyme-catalyzed reaction?

[O & C]

10. What substances form structural elements in living organisms? What properties must a material possess to be suitable to function as a structural element? For any *one* structural component of living organisms show how it fulfils its function.

11. List *four* functions performed by proteins in living organisms, and for any one of these describe how a particular protein carries out its function.

12. What chemical and physical properties are advantageous in storage materials in animals and plants? Illustrate this by reference to any one type of storage material.

13. Cyanide ions inhibit enzymes which have a prosthetic group containing iron.
    (*a*) (i) Define the term prosthetic group.
         (ii) Name *one* enzyme which has iron as a prosthetic group.
         (iii) Indicate the role of this enzyme in cellular metabolism.
    (*b*) (i) How would you determine the rate of a named enzyme-catalysed reaction? (Give experimental details.)
    (*c*) (i) Explain how the use of a named isotope has increased our understanding of any one biological problem.
         (ii) Briefly describe how it was used and monitored under experimental conditions.

[JMB]

Further reading

Holloway, M. R. (1976), 'The Mechanism of Enzyme Action', *Oxford/Carolina Biology Reader*, **45**.

Koshland, D. E. (October 1973), 'Protein Shape and Control', *Scientific American*, **229**, 52.

Maclean, N. (1978), 'Haemoglobin', *Studies in Biology*, **93**, (Edward Arnold).

Perutz, M. F. (November 1964), 'The Hemoglobin Molecule', *Scientific American*, **221**, 64.

Phelps, C. F. (1972), 'Polysaccharides', *Oxford/Carolina Biology Reader*, **27**.

Phillips, D. C. (November 1966), 'The Three-Dimensional Structure of an Enzyme Molecule', *Scientific American*, **215**, 78

Phillips, D. C., and North, A. C. T. (1978), 'Protein Structure', *Oxford/Carolina Biology Reader*, **34**, (Second Edition).

Sharon, N. (May 1974), 'The Bacterial Cell Wall', *Scientific American*, **230**, 78.

Sharon, N. (November 1980), 'Carbohydrates', *Scientific American*, **243**, 80.

Steward, M. W. (1974), 'Immunochemistry', *Outline Studies in Biology*. (Chapman & Hall).

Stroud, R. M. (July 1974), 'A Family of Protein-Cutting Enzymes', *Scientific American*, **231**, 74.

Woodhead-Galloway, J. (1980), 'Collagen: The Anatomy of a Protein', *Studies in Biology*, **117**. (Edward Arnold).

Wynn, C. H. (1979), 'The Structure and Function of Enzymes', *Studies in Biology*, **42**. (Edward Arnold).

3
Obtaining
Energy

Summary

The sequences of reactions by means of which living organisms perform all their various life processes are collectively referred to as 'metabolism'. In addition to raw materials for growth and maintenance, all organisms must obtain a supply of energy with which to drive their metabolic activities. All organisms on earth are ultimately dependant upon the energy of sunlight being trapped by photosynthetic organisms. Trapped light energy is transformed into biologically-usable forms of energy, usually ATP and reducing power. All non-photosynthetic organisms take in highly-reduced forms of carbon (carbohydrates, fats, proteins) and oxidize these to carbon dioxide and water, with the production of energy, again in the form of ATP. The ability to consume a wide variety of foodstuffs demands a versatile metabolic system capable of dealing with an enormous range of substances. Food materials are broken down bit-by-bit to yield a very few suitable compounds which are oxidized in a final common oxidative sequence of reactions. Because many thousands of reactions are potentially capable of occurring in any given cell, there must be control. This is achieved by feed-back mechanisms and by hormones, acting on individual enzyme-catalyzed steps in metabolism.

3.1. Introduction

Energy is a word frequently in the headlines these days, more often than not in the context of 'The Energy Crisis'. Twentieth-century industrial man exploits his environment to obtain energy. Fossil fuels (coal, oil, gas), our major sources of energy, are burnt in steam, petrol and diesel engines to supply power for man's activities. These engines are *heat engines*. In a heat engine, heat must be able to flow from a high temperature region to a low temperature region: the steeper the temperature gradient, the more efficient the engine.

Living organisms also have to obtain energy from the environment in order to survive. Animals obtain energy from food materials and green plants trap light energy from the sun: both convert the energy into a form suitable for driving energy-*requiring* life activities. Living organisms are not heat engines, however: they operate at essentially constant temperature and pressure. They do, of course, experience *changes* of temperature, but they do not *use* temperature differences to drive energy-requiring processes. Large temperature differences are in any case not allowable because many of the molecules of life, as well as delicate structures such as membranes, are disorganized by temperatures only a little above the normal physiological ones. Relatively recently man himself has developed 'non-heat' engines such as fuel cells and solar cells. These convert chemical or light energy directly into electricity (Figure 3.1).

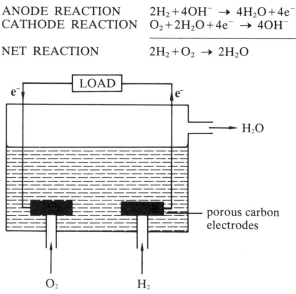

ANODE REACTION $2H_2 + 4OH^- \rightarrow 4H_2O + 4e^-$
CATHODE REACTION $O_2 + 2H_2O + 4e^- \rightarrow 4OH^-$

NET REACTION $2H_2 + O_2 \rightarrow 2H_2O$

Figure 3.1 Diagram of a fuel cell: hydrogen is oxidized and an electric current is produced

Organisms couple energy-releasing processes to energy-requiring ones

Energy is the capacity to do work and appears in many different, interconvertible forms: light, heat, electrical, mechanical, chemical, and so on. Energy can be kinetic ('active') or potential ('stored'). Water behind a dam represents a store of potential energy that could be released to give the kinetic energy of motion.

Everyday experience tells us that all processes may be divided into two categories: those that tend to go by themselves and which may be thought of as energy-releasing, and those that do not, unless energy is expended. Water falling downhill is an example of the first class of process and pulling a vehicle uphill is an example of the second. Two such processes could be *coupled* by using a length of rope (Figure 3.2) so that the energy-releasing process drives the energy-requiring one.

Living organisms also couple energy-releasing processes to energy-requiring ones and a major aim of this chapter is to understand how they do this. Energy-releasing reactions include oxidation of food materials such as carbohydrates and fats. Energy-requiring reactions include muscular contraction and building up protein from individual amino acids. Living organisms can be said to exchange *chemical energy* between energy-releasing and energy-requiring processes. We can begin to understand how they do this by thinking a little about equilibria.

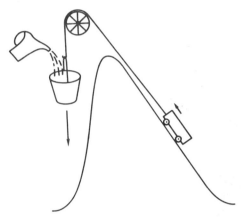

Figure 3.2 Coupling two processes. The water will tend to flow downhill but the truck has no tendency to move uphill. The two processes can be coupled with a piece of rope

Chemical reactions tend to proceed to equilibrium

Energy changes in molecules occur when bonds are broken and new bonds form. If the total energy of the products of a reaction is less than that of the reactants then energy is released by the reaction taking place. The energy of interest to us here is the *free energy* which is the energy available to do work under conditions of constant temperature and pressure. A reaction in which free energy is released is called an *exergonic* reaction. All spontaneous reactions are exergonic, that is they proceed 'downhill' with the release of energy. To put it another way, all systems tend to minimize their free energy content.

We can consider every molecule to contain an amount of potential energy equal to the amount necessary to synthesize it originally. Thus a living organism made up of organic compounds is a store of potential energy. Breaking down these organic compounds will release free energy. However, it is found that every energy transformation results in a reduction in the amount of usable, or free, energy of the system because of the increasing amount of disorder. It is therefore not possible to use all the energy released in a reaction to perform work: some may be transformed into heat or dissipated in increased disorder in the system.

A strongly exergonic reaction will tend to proceed to completion. If, on the other hand, the energy difference between reactants and products is not very great, there will be some tendency for the reaction to go in either direction—although it must still go 'downhill'. Such a reaction is described as an *equilibrium*. In contrast to this, energy-consuming reactions, called *endergonic reactions*, do not, by themselves, proceed. However, they can be made to proceed by incorporating them into an overall process which is, in total, exergonic.

Suppose there is an equilibrium reaction by which A reacts with B to give products C and D:

$$A+B \rightleftharpoons C+D$$

Experience tells us that reactions proceed in such a way as to change the relative concentrations of reactants and products towards an equilibrium value—the point at which the free energy is minimized. The equilibrium concentrations are expressed as a ratio, the equilibrium constant, K_{eq}:

$$K_{eq} = \frac{[C][D]}{[A][B]}$$

where the square bracket means 'concentration of'. For reactions where the concentrations of C and D at equilibrium are greater than those of A and B, K_{eq} is greater than 1.0, and *vice versa*. If we know K_{eq} for a given reaction as well as the initial concentrations of the reactants, we can predict the direction in which chemical change will occur. Clearly, the farther away the initial concentrations are from the equilibrium ones, the more potential there is for the reaction to proceed. Furthermore, if we raise the concentration of reactants, or remove products, the reaction can be made to proceed. This is the *Law of Mass Action*: adding A and B to an equilibrium mixture of A, B, C and D results in the conversion of some A and B into C and D. Adding C and D has the opposite effect. Now we can see how it is possible to drive reactions having little or no tendency to proceed (i.e. endergonic reactions, those with K_{eq} less than 1.0). We can *couple* them to an exergonic reaction by means of a common intermediate. Suppose we have two reactions:

$$A+B \rightleftharpoons C+D \quad \text{(exergonic: } K_{eq} > 1.0) \tag{1}$$

$$D+X \rightleftharpoons Y+Z \quad \text{(endergonic: } K_{eq} < 1.0) \tag{2}$$

Starting with A and B at concentrations, relative to those of C and D, greater than their equilibrium concentration, will result in the formation of some C and D. Raising the concentration of D causes reaction (2) to proceed to re-establish equilibrium. Compound D is the *common intermediate*. It is possible to show that the free energy change, under standard conditions, is directly related to the equilibrium constant.

Two words of caution are necessary at this point. Firstly, although ideas about equilibrium are helpful in aiding our understanding of biological pro-cesses, it is important to realize that the condition of equilibrium is rarely reached in living organisms. Indeed the driving force for reactions is measured by how far the concentrations of reactants are *displaced* from the equilibrium ones. Once equilibrium is reached there is no driving force. Usually living organisms operate under *steady-state conditions*, not equilibrium ones. We can visualize this by analogy with a bathtub filled with water as far as the overflow and with the taps running. As long as water continues to be added it comes out of the overflow; however, the level (i.e. the concentration) of the water in the bath does not change (Figure 3.3). Typically, reactions in biological systems go in sequences:

compound 1 → compound 2 → compound 3 → compound 4, and so on.

None of the individual reactions is ever at equilibrium. Using up compound 4 in some further reaction causes the conversion of more compound 3 to compound 4, and this causes more compound 2 to turn into compound 3, and so it goes on.

Secondly, in the above discussion of energy and equilibrium we said nothing about how *fast* the reactions would go. The fact that there is a large amount of stored potential energy does not mean that the release of energy will be rapid. A large body of water behind a high dam may be released explosively or as a trickle. The rate of release of energy depends upon the activation energy for that reaction and whether or not catalysts are present (see p. 64).

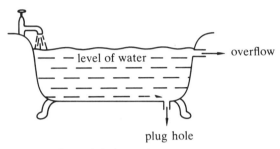

Figure 3.3 A steady-state system

3.2. Metabolism

Armed with this basic information about energy we can now go on to look at the reactions proceeding in living cells. In particular, we will be keeping an eye open for how exergonic reactions are coupled to endergonic ones.

The sum total of all the reactions going on inside a living cell is called *metabolism*. The compounds participating in these reactions are sometimes called *metabolic intermediates* or *metabolites*. Of course, they are also substrates and products of the various enzymes catalyzing the reactions. In any given cell many thousands of metabolic reactions are going on simultaneously, each being catalyzed by an individual, specific enzyme.

Metabolism is divided, mainly for convenience, into *catabolism* and *anabolism*, meaning breaking down and building up, respectively. Obviously, catabolic reactions will tend to be exergonic and anabolic reactions endergonic. A large part of biochemical research is concerned with finding out about metabolism (Box 3.1).

Catabolic reactions are those in which molecules in food materials or in stores (starch, glycogen, fat) are broken down. Such reactions have a dual purpose. They produce the raw materials or building blocks for building up new cellular substance and they also provide the energy for doing this, as well as providing energy for muscular work, etc. Anabolic reactions are those in which complex molecules, including macromolecules, are built up from simpler molecules. These are *biosynthetic reactions*. In photosynthetic organisms the energy of sunlight is used to drive biosynthetic reactions in which the very simple molecules taken up by plants (carbon dioxide, water, ammonium ions) are converted to the sugars, fats and proteins that form plant tissues. In the dark, plants depend for energy on the oxidative catabolism of stored materials such as starch (Figure 3.4).

Box 3.1 Finding out about metabolism

There are so many different reactions going on simultaneously in a living cell that a biochemist has to be something of a detective to find out what is happening. Thousands of enzymes and thousands of substrates all doing the right things at the right times! The main aim of the biochemist is to discover a *metabolic pathway*, that is to discover the sequence of enzyme-catalyzed steps by which one compound is converted to another. Such pathways may be very complicated: complete oxidation of glucose to carbon dioxide, for example, involves over thirty enzyme-catalyzed steps. However, each step is fairly simple and involves only a small change to the molecule involved. You might compare this with the word game.

$$
\begin{array}{ll}
\text{FOOD} & \text{GLUCOSE} \\
\text{FOOT} & \big| \\
\text{FORT} & \\
\text{PORT} & \big\downarrow \\
\text{PART} & CO_2 + H_2O
\end{array}
$$

Therefore one can often predict what 'the next step' is likely to be. But where to start? There is no single way to start solving the puzzle and very many advances in biochemical knowledge have depended on a biochemist having an ingenious idea. Nevertheless we can mention here three general ways by which metabolic puzzles can be solved. These are:

(1) Make the system as simple as possible.
(2) Trace what happens to a compound by labelling it.
(3) Use 'poisons' to 'stop' the system at some intermediate point.

We will look at these briefly in turn here.

(1) Make the system as simple as possible. Instead of using a whole animal, use an isolated, but still 'living' organ or tissue (liver, kidney, brain). Instead of using whole cells, use organelles (mitochondria, nuclei, chloroplasts) isolated from them: these organelles have their own specific functions. Thus, to study the reactions of photosynthesis use isolated chloroplasts.

(2) Trace what happens to a compound by labelling it. It would be nice to put some sort of tag or label on a compound so that its metabolites can be identified. A chemical 'label' can sometimes be used, but chemical tagging has the disadvantage of changing the properties of a compound so that it may not be 'handled' by the enzymes of the pathway: they may simply fail to recognize it. It is much more satisfactory to label the compound by replacing one or more of its atoms by an isotope of the same element. This will have a slightly different mass but identical chemical properties. Although the metabolic system will not be able to tell the difference between the isotopically-labelled compound and the natural one, *we* shall be able to, using physical methods. Isotopes can be distinguished by their mass using an instrument called a mass spectrometer. Since 1946, however, radioactive isotopes have been available to biochemists, and are more commonly used in metabolic studies. Isotopes available include those shown in the table below.

Element	Non-radioactive	Radioactive
Hydrogen (^{1}H)	^{2}H	^{3}H
Carbon (^{12}C)	^{13}C	^{14}C
Nitrogen (^{14}N)	^{15}N	–
Oxygen (^{16}O)	^{18}O	–
Phosphorus (^{31}P)	–	^{32}P

Radioactive isotopes can be detected easily because of their emissions by using photographic film, a Geiger counter or a scintillation counter. Tiny amounts can be detected with great accuracy. The uptake of $^{14}CO_2$ by a plant in the course of photosynthesis could be followed. If radioactively-labelled starch was later obtained from the plant this would show that CO_2 is converted to starch. This technique is of very great importance to the biochemist in tracing one compound amongst many: imagine how easy it would be to find a radioactive needle in a haystack if one had a Geiger counter!

(3) Use 'poison' to 'stop' the system at some intermediate point. Many poisons are in fact inhibitors of enzymes. Adding small amounts of a poison to a metabolic system may inhibit just one enzyme of a pathway. This is like putting a 'spanner in the works'; compounds before the blockage accumulate. Imagine a sequence of metabolic reactions:

$$A \rightarrow B \rightarrow C \overset{\text{poison}}{\longrightarrow\!\!\!|\!\!\!\longrightarrow} D \rightarrow E$$

Adding a metabolic poison may inhibit the enzyme that converts compound C to compound D. Compound C would normally be present in extremely tiny amounts, but the blockage will cause it to accumulate. It may then be isolated and its structure analyzed. This will tell the biochemist that the pathway from A to E goes via C. He can then do other experiments in the hope of finding out the nature of the other metabolic intermediates in the pathway.

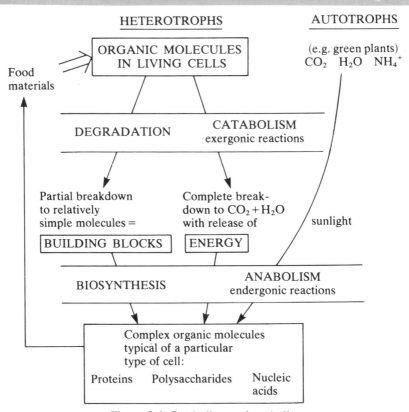

Figure 3.4 Catabolism and anabolism

Energy currency

A large amount of energy is released when a sugar such as glucose is completely oxidized to carbon dioxide and water. A large amount of energy is needed to couple several hundred amino acids together to make a protein. Neither of these reactions can be looked upon as an equilibrium reaction. The first, to all intents and purposes, would go all the way to completion, the second will not go at all. How can the two processes be coupled together? What is the common intermediate? The answer can be expressed in three letters: ATP.

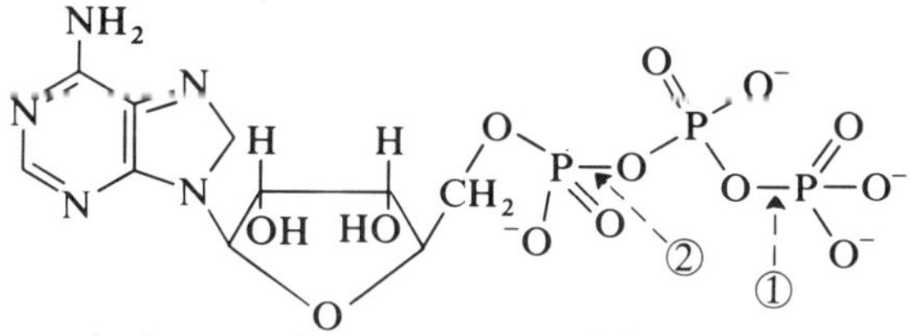

Hydrolysis at ① yields adenosine diphosphate (ADP) and orthophosphoric acid (H_3PO_4, usually written P_i to represent 'inorganic phosphate'). Hydrolysis at ② yields adenosine monophosphate (AMP) and inorganic pyrophosphate ('PP_i')

Figure 3.5 Structure of adenosine triphosphate (ATP)

ATP is *a*denosine *tri-p*hosphate, a *nucleotide* compound formed by linking a nitrogen-containing base, adenine, to a pentose sugar, ribose, which in turn is linked to a string of three phosphate groups (Figure 3.5). ATP can act as an intermediary or common intermediate by virtue of the fact that the removal of a phosphate grouping by hydrolysis is a highly-exergonic reaction (Box 3.2): the products of hydrolysis are the diphosphate, adenosine diphosphate or ADP, and 'inorganic phosphate', P_i , actually orthophosphate:

$$\text{adenosine triphosphate} + H_2O \rightleftharpoons \text{adenosine diphosphate} + H_3PO_4$$

(ATP)   (ADP)   (P_i)

This reaction has a strong tendency to go from left to right because the equilibrium constant is high, but reaction may take place at an appreciable rate only in the presence of the appropriate catalysts. As we shall see, ATP formation—that is the above reaction going from right to left—occurs in photosynthesis and during catabolism of foodstuffs. The reaction in this direction will not proceed spontaneously, but can be made to go if it is coupled with a highly-exergonic process so that the net reaction is also exergonic. In anabolic reactions the opposite occurs: ATP hydrolysis has a strong tendency to occur and may be coupled with endergonic reactions of biosynthesis such that the overall reaction is exergonic. The secret lies in organizing both energy-yielding and energy-requiring reaction sequences so that ATP formation or ATP hydrolysis takes place during the course of the sequences.

Because of its function in linking highly-exergonic with highly-endergonic processes, ATP is a reservoir of potential chemical energy. In order to emphasize its role in energy metabolism it has been called a 'high-energy compound' or is said to possess 'high-energy bonds'. Although these terms serve to remind us of the function of ATP, we should remember that there

Box 3.2 Adenosine triphosphate (ATP)

The particular importance of ATP in biological systems lies in its string of three phosphate groups. One of these is connected to the ribose via an ester linkage, but the other two are phosphates coupled together. Phosphates linked in this way are called *pyrophosphates* and form as a result of the elimination of water between two phosphoric acid units. We can compare such bonds with acid anhydrides:

$$
\begin{array}{ccc}
\text{acetic anhydride} & \xrightleftharpoons[\;]{H_2O} & \text{acetic acid} \\
\text{pyrophosphate} & \xrightleftharpoons[\;]{H_2O} & \text{phosphoric acid}
\end{array}
$$

Such compounds in water at neutral pH are easily hydrolyzed. In the case of ATP we can write the process in two steps because there are two pyrophosphate bonds.

$$
\underset{\text{ATP}}{A-P-P-P} \xrightarrow{H_2O} \underset{\substack{\text{adenosine} \\ \text{diphosphate} \\ \text{(ADP)}}}{A-P-P+P_i} \xrightarrow{H_2O} \underset{\substack{\text{adenosine} \\ \text{monophosphate} \\ \text{(AMP)}}}{A-P+P_i}
$$

P_i means 'inorganic phosphate' or phosphoric acid. The first cleavage is especially important in biological systems although both steps do occur on occasions.

The hydrolyses described have high equilibrium constants. For the reaction:

$$ATP + H_2O \rightleftharpoons ADP + P_i$$

K_{eq} is of the order of 10^7 and therefore at equilibrium there is very much more ADP and inorganic phosphate present than ATP. From our knowledge of the equilibrium constant, or alternatively from the standard free energy change for the reaction which is high and negative, we can predict that the hydrolysis of ATP is an exergonic reaction which will tend to proceed spontaneously. Of course, it is *potential* energy we are talking about and the reaction may not proceed at a measurable rate unless a catalyst is present. ATP is in fact reasonably stable under ordinary conditions. Like most other compounds found in living organisms it is *thermodynamically unstable but kinetically stable*.

is nothing unusual about ATP as a chemical substance or about its bonds. It is merely a compound with a strong tendency to hydrolyze. One way of looking at this is to consider ATP as a *dehydrating agent*. It will tend to 'pull' water from other reactions and become hydrolyzed in the process. Thus in the biosynthesis of proteins, it is required that amino acids link together in a condensation reaction with the elimination of water, to form a peptide bond (see Figure 2.8). ATP is used in protein biosynthesis, although the actual reactions involved are very much more complicated.

ATP plays many roles in the cell. In addition to supplying the driving force for biosynthetic reactions, ATP hydrolysis can supply the energy for muscular

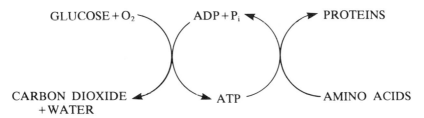

Reactions such as the oxidation of food-stuffs are coupled to the manufacture of ATP from ADP and P_i; synthetic, energy-requiring reactions such as building protein molecules from amino acids are coupled to the hydrolysis of ATP to ADP and P_i (or to AMP and PP_i in some instances).

ATP hydrolysis supplies the driving force for practically all the energy-requiring processes performed by living organisms

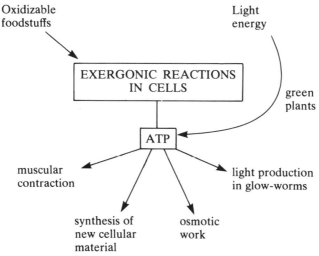

Figure 3.6 Energy-yielding processes may be coupled to energy-requiring processes through the ATP–ADP cycle

contraction (p. 169), for the performance of osmotic work and even for light production in glow-worms (Figure 3.6). Many of these reactions would not be practicable in free solution—but then the reactions of living organisms do not go on in free solution but rather occur on the surfaces of highly-specific enzyme catalysts.

ATP can be thought of as acting as a currency in the cell and the analogy is a useful one. Currencies, like paper money, have no intrinsic value, but are accepted as an improvement and a simplification of the system of barter. ATP is not stored (stored currency might be stolen or succumb to the ravages of inflation, and attracts no interest) and is not transported between cells (currency may be stolen or lost in transit). If there is excess energy-yielding capacity, such as just after a meal, then the ATP is 'invested' by using it to drive the synthesis of storage materials such as fat and glycogen. The 'interest' could be thought of as having fat as a thermal and mechanical insulator (Figure 3.7).

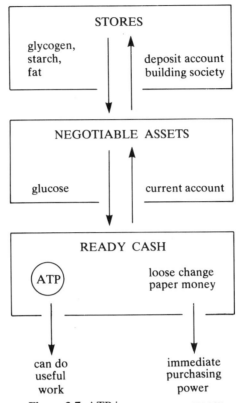

Figure 3.7 ATP is an energy currency

Thus the only remarkable property of ATP is its ability to participate in such a multitude of reactions. Why then was ATP selected? Would not simple pyrophosphate, for which the equilibrium constant for hydrolysis is about 1.4×10^6, be just as good? In energy terms it would be, but the fact that ATP has an adenine and a ribose group, equivalent to a 'handle', probably makes it possible for enzymes to recognize it with a high degree of specificity.

Both ATP and pyrophosphate have high equilibrium constants for their hydrolysis at neutral pH because the products of reaction have negative charges. To perform the reverse reaction requires that these molecules with like charges come together. Since like charges repel, this is a comparatively unlikely event. To put it another way, it costs energy to bring them together.

Most of the rest of this chapter will be concerned with understanding how ATP is produced in photosynthesis and during catabolic reactions while Chapter 4 will be concerned with how ATP is used.

Where is ATP produced?

Most (but not all) of the ATP production in all cells goes on in specialized membranous structures. These are the mitochondria (in animals and plants), the chloroplasts (in green plants) and the plasma membrane (in prokaryotes). The fact that membranous structures are involved has made this process very difficult to study. Although biochemists are quite good at investigating the reactions involving soluble enzyme proteins, they are not so good at dealing with insoluble systems where enzymes and cofactors are embedded in lipid-containing membranes. Where it has been possible to extract enzymes from the membrane for study, there is no guarantee that these enzymes behave in isolation as they do when they are inserted in the membrane. Some of the enzymes are impossible to extract in an active form. Much work has been done with 'smashed up' mitochondria and chloroplasts, that is with *fragments* of these organelles. Although such investigations have yielded many important experimental findings it is rather like smashing a wrist-watch with a sledge-hammer, and then examining the bits and trying to predict how the watch worked.

Box 3.3 Oxidation and reduction

In its simplest terms reduction means the addition of an electron, e^-, to a compound, and oxidation means the removal of an electron. Removal of electrons from compounds releases energy: to replace the electron obviously costs energy.

In chemical reactions, electrons are passed from one compound to another. A compound that can donate electrons is called a reducing agent or an electron donor: the recipient of the electrons, an electron acceptor, is an oxidizing agent. Whenever one compound is reduced, another compound must become oxidized:

$$Ae^- \quad + \quad B \longrightarrow A \quad + \quad Be^-$$

| donor | acceptor | A is more 'oxidized' | B has been 'reduced' |

A compound may be reduced in a number of ways, but very commonly reduction involves addition of electrons plus hydrogen atoms, for example:

$$AH_2 + B \rightarrow A + BH_2$$

Similarly, oxygen atoms can be added during an oxidation:

$$A + BO \rightarrow AO + B$$

In general, the process of removing hydrogen atoms or of adding oxygen atoms is referred to as an oxidation because of the different levels of oxidation through which an element may pass:

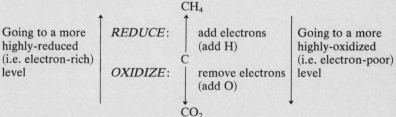

| Going to a more highly-reduced (i.e. electron-rich) level | *REDUCE*: | add electrons (add H) | Going to a more highly-oxidized (i.e. electron-poor) level |
| | *OXIDIZE*: | C remove electrons (add O) | |

Although hydrogen and oxygen atoms are involved it is more important to concentrate on the electrons.

Oxidation–reduction: electrons and protons

Chemical reactions are concerned with electrons, and when a chemical change takes place electrons shift to new configurations. If the available free energy of the first configuration is higher than that of the new one, then energy is released, and *vice versa*. Living organisms take electrons at a high-energy potential and drop them to a 'sink' at a lower energy potential (Box 3.3). This process is coupled to the energy-requiring synthesis of other compounds.

With a few exceptions there is only one primary process on earth that provides a source of electrons at a high potential—*photosynthesis*: all other life processes are dependent on this. In photosynthesis the absorption of light energy results in electrons being promoted to a higher energy state and the energy is trapped as chemical energy as the electrons fall back to their original state. Carbohydrates such as starch and fats, in food and as stores in both animals and plants, are also sources of electrons at a high potential. However, these substances are all originally built up by plants using the energy derived from photosynthesis, so that they are really *secondary* sources of energy. Animals obtain them by eating plants or other animals.

What sorts of compound can act as 'sinks' of lower potential? Oxidizing agents obviously, and we could think of many compounds that could serve this function. A very simple example is the change from iron (III) to iron (II), i.e. $Fe^{3+} + e^- \rightarrow Fe^{2+}$. However, molecular oxygen is the compound with almost unrivalled ability to act as an electron acceptor (i.e. as an oxidizing agent), and in metabolism, overall reactions of the type:

$$\text{substrate--H}_2 \quad + \tfrac{1}{2}O_2 \rightarrow \text{substrate} + H_2O$$

are the most important source of energy for cellular processes. Reactions like this go on in the mitochondria of both animal and plant cells: the reactions are coupled to the formation of ATP from ADP. Since photosynthesis is the primary energy-conserving process on earth we will deal with it first.

3.3. Photosynthesis

The overall process of photosynthesis involves both the absorption of light energy and the use of this energy to build complex plant materials from simple molecules like water, carbon dioxide and ammonium ions. Sugars, that is carbohydrates, are often thought of as a major product of photosynthetic activity and so the process is often written:

$$6CO_2 + 6H_2O \xrightarrow{\text{light energy}} C_6H_{12}O_6 + 6O_2$$

This equation simplifies, indeed oversimplifies, a complicated process made up of very many steps. Can we disentangle a few simple principles in order to understand what is going on? Remembering that carbohydrates are energy-rich and represent a highly-reduced form of carbon, while carbon dioxide is energy-poor and highly oxidized, let us start by dividing the above equation by 6:

$$CO_2 + H_2O \rightarrow (CH_2O) + O_2$$

This shows that a highly-oxidized form of carbon is being *reduced* to the level at which it occurs in carbohydrates. In order to do this *reducing power* is required: apparently water is acting as the reducing agent supplying protons and electrons. Because carbon atoms are being joined together to make carbohydrate, i.e.

$$6(CH_2O) \rightarrow C_6H_{12}O_6$$

a supply of *energy* is required. By now we ought not to be surprised to find that this comes from ATP.

Two experiments were of great historical importance in the understanding of photosynthesis. In 1937, Hill found that the generation of reducing power could be completely divorced from the reduction of carbon dioxide when he showed that illuminated chloroplasts would reduce certain dyes (the so-called Hill Reaction). This demonstrated that 'reducing power' is produced in light. Then in 1954 Arnon and his co-workers showed that illuminated chloroplasts could produce ATP—again a process completely divorced from the reduction of carbon dioxide to carbohydrate. The process of photosynthesis can therefore be written as two overall steps (Figure 3.8). It is, in fact, possible to separate these processes, that is to say, *given a supply of reducing power and ATP, light is not required for carbohydrate synthesis in plants.* Only the first step, the *light stage* of photosynthesis, interests us here: the second step, or *dark stage*, will be considered in Chapter 4.

STEP I *Light required*
Light energy trapped, i.e. electrons raised to higher potential used to generate: (a) reducing power, and
(b) biologically-usable energy = ATP

STEP II *Light not required*
Reducing power and ATP used in converting highly-oxidized form of carbon (CO_2) to highly-reduced forms of carbon such as carbohydrates, fats, and proteins—indeed all the compounds of which plants are composed

Figure 3.8 Photosynthesis can be divided into two overall steps, a light reaction and a dark reaction

The light stage of photosynthesis

We tend to associate photosynthetic activity with green plants, but in fact it also goes on in the blue-green algae (Cyanophyta) and certain bacteria. It has been estimated that up to half of the photosynthesis on earth takes place in these unicellular organisms. It was originally thought that this was an entirely different process, partly because photosynthetic bacteria do not evolve oxygen. However, when it was realized that photosynthetic bacteria use compounds other than water to supply the protons and electrons required to reduce carbon dioxide it became clear that the process had a lot in common with that occurring in green plants. Bacteria use a range of different compounds, both organic and inorganic, as proton and electron donors: an example is hydrogen sulphide. The equation for this process can be written:

$$2H_2S + CO_2 \rightarrow (CH_2O) + H_2O + 2S$$

Here, elemental sulphur is produced rather than molecular oxygen.

Going back to the equation for green plant photosynthesis, we may now write:

$$2H_2O + CO_2 \rightarrow (CH_2O) + H_2O + O_2$$

Of course, we *could* cancel out a water molecule on either side, but experiments using different isotopes of oxygen show that we should not do this because *all* the oxygen evolved comes from water. Two water molecules are required to provide one oxygen molecule (Figure 3.9).

$$2H_2{}^{18}O + C^{16}O_2 \rightarrow (CH_2{}^{16}O) + H_2{}^{16}O + {}^{18}O_2$$
$$\uparrow \qquad\qquad\qquad\qquad\qquad\qquad \uparrow$$

Figure 3.9 Where does the oxygen produced in green plant photosynthesis come from? This problem can be solved by using the isotope of oxygen, $^{18}O_2$. This heavy isotope is not radioactive and has to be detected and measured by mass spectrometry

We can therefore devise a general equation for all types of photosynthesis, writing a hypothetical proton and electron donor as H_2A:

$$2H_2A + CO_2 \rightarrow (CH_2O) + H_2O + 2A$$

H_2A is split to produce reducing and oxidizing fragments, and the oxidizing fragment is liberated (e.g. oxygen, sulphur or some other compound).

What is 'reducing power'?

'Reducing power' for photosynthesis means a reducing agent capable of supplying protons and electrons to reduce carbon dioxide to organic compounds. This reducing agent must be able to participate in the enzyme reactions in which these organic compounds are produced. In photosynthesis the reducing agent involved is a coenzyme with a long name, nicotinamide adenine dinucleotide phosphate, that can exist in either oxidized or reduced form, abbreviated $NADP^+$ or NADPH (Figure 3.10). Although this molecule has a complicated structure, we can think of it as having a 'business end' which can carry protons and electrons, and a 'handle' which enables it to be recognized with a high degree of specificity by the appropriate enzymes. We will see later (Chapter 4) that when reducing power is needed in biosynthetic reactions in living organisms (not just those of photosynthesis), NADPH is almost always the coenzyme involved.

Where does photosynthesis take place?

In eukaryotic cells photosynthesis takes place in the specialized organelles, the chloroplasts, found in the mesophyll cells of leaves and elsewhere. Chloroplasts are bounded by a double membrane and contain a membrane system which is highly convoluted to form stacks of flattened sacs. The stacks are called *grana* and are embedded in a fluid matrix called the *stroma* (Figure 3.11). The chloroplasts carry out the whole of photosynthesis (both light and dark reactions), but the light reaction goes on solely in the grana. Photosynthetic bacteria do not have chloroplasts, but nonetheless do have a membrane system that performs photosynthesis.

A *nucleotide* is a compound with the structure:

NITROGEN-CONTAINING BASE—SUGAR—PHOSPHATE

The outline structure shown below consists of two such compounds joined phosphate-to-phosphate, and is therefore a *di-*nucleotide. (It happens to have an *extra* phosphate on one of the sugars, hence NADP^+.)

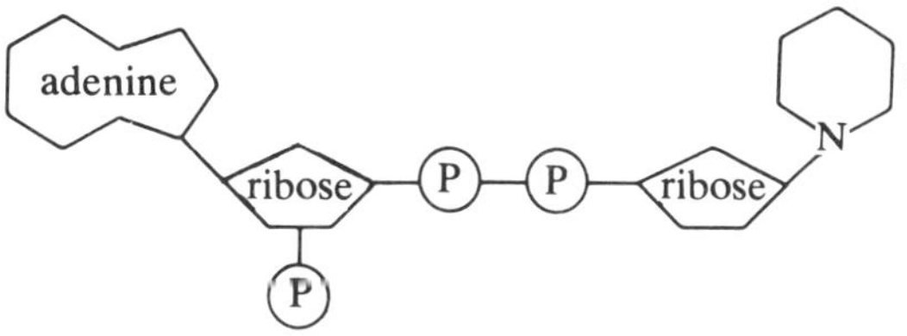

This is a coenzyme whose function is to carry protons and electrons. Although its structure is complex, we can think of it as having a 'business end' (the nicotinamide ring) that does the carrying, and a 'handle' (all the rest) which is capable of being recognized by an enzyme.

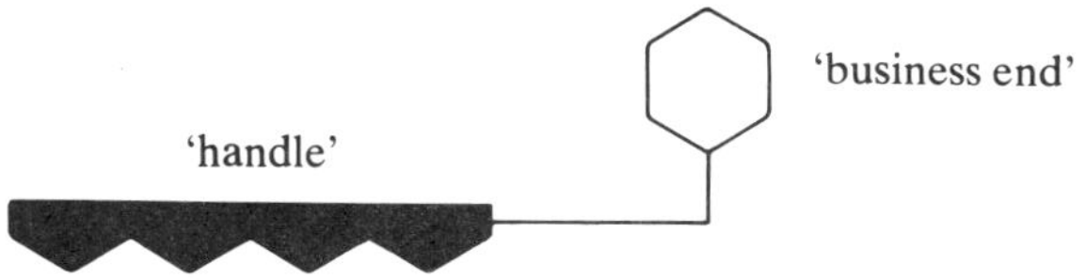

The 'business end' or nicotinamide ring, can carry, reversibly, one proton and two electrons. Since oxidation reactions involve the removal of equal numbers of protons and electrons (usually $2H^+$ and $2e^-$) from a compound, it is usual to write an additional proton on the right-hand side of the equation. This makes the total charge on each side equal since the N of the nicotinamide ring already carries a positive charge.

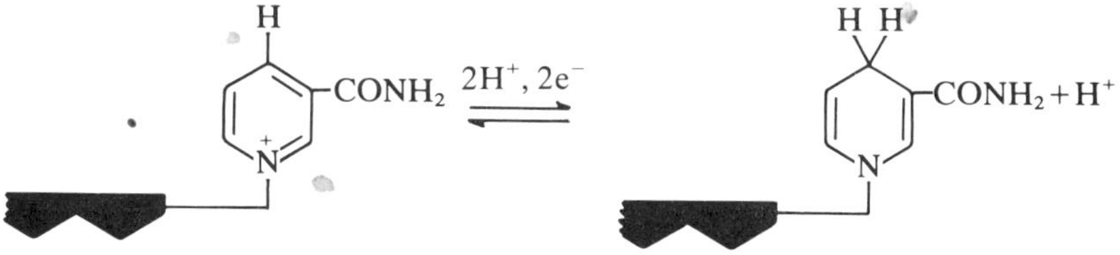

This reaction is usually written:

$$NADP^+ + 2H^+ + 2e^- \rightleftharpoons NADPH + H^+$$

(oxidized   (reduced
form)   form)

Figure 3.10 'Reducing power' is reduced $NADP^+$. $NADP^+$ stands for nicotinamide adenine dinucleotide phosphate.

How is light energy trapped?

When light falls on an object it may pass straight through, that is, be transmitted, or be reflected or be absorbed. Molecules only tend to absorb light of certain characteristic wavelengths. If molecules of a compound appear to be green to our eyes when illuminated with white light they are absorbing red, blue and yellow light, but are not absorbing the green. The absorption *spectrum* of a compound is a record of which wavelengths are absorbed and which are

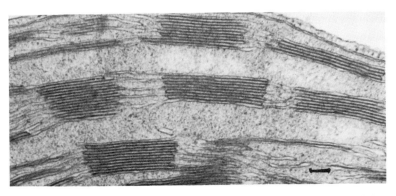

Figure 3.11 Where the reactions of photosynthesis take place. The *thylakoids* are flattened sacs, and a pile of these stacks is called a *granum*. The space outside the grana but inside the chloroplast membrane, and which contains soluble enzymes, is called *stroma*. See also Figs. 1.20 and 1.21. The thylakoids arise from invaginations of the chloroplast inner membrane and are therefore analogous to cristae in mitochondria (see Fig. 1.19). The thylakoids contain chlorophyll and all the other energy-transducing machinery. Stroma contains the soluble enzymes that utilize ATP and NADPH to convert carbon dioxide to organic material (sugar). Length of line 0.1 μm

not. When light is absorbed by a compound, the energy associated with the light is then trapped by that compound. If light energy is to be absorbed in photosynthesis, then one or more light-absorbing compounds or pigments must be present. There is, in fact, a range of pigments in most photosynthetic tissues, but always some form of *chlorophyll* is present (Figure 3.12). Photosynthetic bacteria contain a characteristic 'bacteriochlorophyll'.

When we compare the absorption spectrum of chlorophyll with the wavelengths of light that are most effective in driving photosynthesis, the so-called *action spectrum* (Figure 3.13), we find that they match reasonably well, but not exactly. Light of intermediate wavelengths between the two major chlorophyll absorbances is more effective than would be predicted. This is because other pigments, such as the carotenoids, can absorb light at the intermediate wavelengths to some extent and can pass the absorbed energy on to chlorophyll. These are called accessory pigments and obviously enable the plants to trap light at more wavelengths than would be possible if they possessed chlorophyll alone.

The absorption of light by a chlorophyll molecule results in the chlorophyll acquiring energy. This energy is transferred in some way to an electron in the pigment molecule. This electron is raised to a higher, relatively unstable energy level. In this state it could drop back to its original low-energy state, re-emitting the light energy absorbed in the process. This happens with chlorophyll in free solution and fluorescence occurs. However, when the chlorophyll molecule is properly arranged within the chloroplast membrane the high-energy electron leaves the chlorophyll molecule and is passed on to an acceptor molecule. If we imagine that the acceptor molecule is $NADP^+$, the result would be to generate reduced $NADP^+$ or reducing power! This does not happen directly, but is the end result of the series of reactions that do occur.

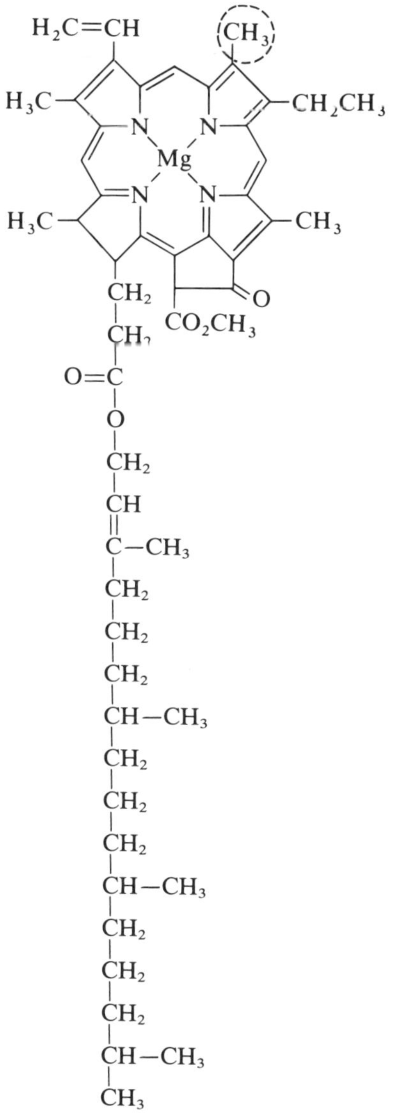

Figure 3.12 The chlorophyll molecule consists of a magnesium porphyrin with a long phytyl tail. Chlorophyll *a* is shown: chlorophyll *b* differs only in that the group circled is an aldehyde, —CHO, instead of a methyl group, —CH$_3$

How are the pigment molecules organized within the chloroplast?

It is believed that the chlorophyll molecules, along with the accessory pigments, are organized into groups which act as light-trapping 'antennae' called *photo-synthetic units*. One unit contains between 200 and 400 pigment molecules, each of which can absorb light. However, one of these pigment molecules is thought to be distinct by virtue of being closely associated with protein molecules, and acts as a reaction centre. One photon of light may strike any

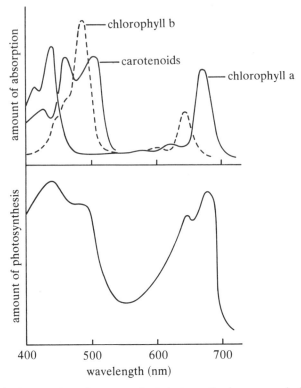

Figure 3.13 Action spectrum for green plant photosynthesis compared with the absorption spectra of chlorophylls and carotenoid pigments. The action spectrum measures the amount of photosynthesis per incident number of photons at different wavelengths. The absorption spectrum measures the amount of light absorbed by a pigment at different wavelengths. The action spectrum (lower curve) can be approximated by summing the absorption curves for chlorophylls *a* and *b* and carotenoids. This suggests that these pigments are involved in light absorption during photosynthesis

pigment molecule within the photosynthetic unit: the energy is then channelled to the reaction centre molecule which is the one that passes on the high-energy electron to the acceptor mentioned above. The organization and orientation of the pigment molecules within the chloroplast is very important, and the same reactions will not proceed in free solution, i.e. with the molecules organized in a random fashion with respect to each other. We could think of a chloroplast as a *solid-state device* for collecting light energy: compare the action of a photoelectric cell (Figure 3.14, Box 3.4).

In green plant chloroplasts there are always two types of photosynthetic units, called photosystems I and II, although both function in rather similar ways. Bacterial photosynthesis, on the other hand, uses only a single photosystem (see below). Photosystem I of green plant chloroplasts is the one that passes electrons eventually to $NADP^+$. In photosystem I, the reaction centre is a molecule of chlorophyll *a* with special properties and which absorbs light of wavelength 700 nm. For this reason it is referred to as P700. The molecule which accepts the high-energy electron from P700 is simply designated as 'X' because its chemical nature is uncertain. Electron-rich, that is to say *reduced*,

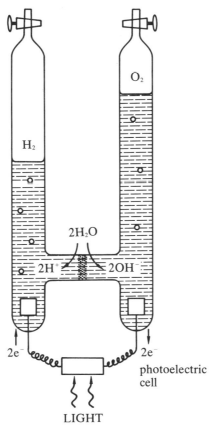

Figure 3.14 An analogy with the production of reducing power and oxygen by photosynthesis is the (admittedly rather impractical) arrangement of a photoelectric cell and electrolysis apparatus shown above. Here the reducing power produced is gaseous hydrogen. The hydrogen and oxygen formed could be recombined to form water with the (explosive) release of the energy originally trapped

Box 3.4 Energy and light

Light is a form of electromagnetic radiation. Visible light forms only a small fraction of the electromagnetic radiation that arrives at the surface of the earth.

wave-
length 10^{-10} 10^{-8} 10^{-6} 10^{-4} 10^{-2} 10^{0} 10^{2} 10^{4}
(nm)

gamma   ultra-   infra-   radio   BBC Radio 4
rays   X- rays   violet   red   waves   (1 500 m)

visible

wave-   400   500   600   700
length
(nm)   BLUE   YELLOW GREEN   RED

There is a relationship between the energy per photon of electromagnetic radiation and wavelength:

$$E = h\nu = \frac{hc}{\lambda}$$

where h is Planck's constant $(6.626 \times 10^{-34}\,\text{J\,s}$, ν is the frequency, λ is the wavelength, and c is the velocity of light $(3 \times 10^8\,\text{m\,s}^{-1})$. The longer the wavelength the smaller the amount of energy.

One photon by itself has very little energy in terms of chemical transformations. We can see this from an example. Normally we use a mole of photons, i.e. Avogadro's number of photons (this is called an *einstein*). For red light of wavelength 700 nm we have:

$$E = \frac{Nhc}{\lambda} = \frac{6.023 \times 10^{23} \times 6.626 \times 10^{-34} \times 3 \times 10^{10}}{7 \times 10^{-5}} = 171\,\text{kJ einstein}^{-1}$$

For comparison, the oxidation of glucose to carbon dioxide and water has a standard free energy change of $-2\,870\,\text{kJ mol}^{-1}$.

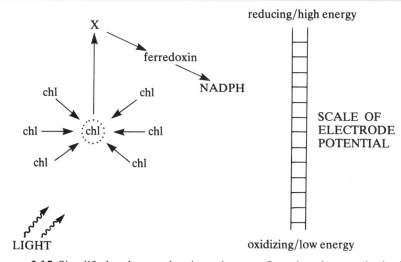

Figure 3.15 Simplified scheme showing electron flow in photosynthesis from chlorophyll to $NADP^+$. Obviously the electrons in the chlorophyll molecules will have to be replenished somehow

X then passes an electron to a second carrier, a small iron-containing protein called ferredoxin. Finally, reduced ferredoxin passes electrons to $NADP^+$ to give reduced NADP: this reduced form picks up protons to form NADPH, and hence reducing power has been generated (Figure 3.15). All of these reactions are catalyzed by specific enzymes in the chloroplast.

Refilling the holes

It should now be obvious that we cannot go on removing electrons from chlorophyll in this way and passing them to $NADP^+$. Everytime a chlorophyll molecule expels an electron a positively-charged 'hole' is left behind which has to be filled up. We therefore need a supply of electrons from somewhere, i.e. we need a reducing agent.

In bacterial photosynthesis many different compounds act as reducing agents. Hydrogen sulphide has already been mentioned. When this compound is 'split', the electrons pass to chlorophyll and eventually to $NADP^+$, while the protons effectively end up in NADPH, and the sulphur is deposited as elemental sulphur. Because hydrogen sulphide is a reasonably powerful reducing agent, only one photosystem is needed.

In green plants, however, the ultimate source of the electrons is water—which is not a particularly good reducing agent. For this reason, two photosystems are used in green plant photosynthesis. The action of light on photosystem I produces NADPH and leaves chlorophyll with a positively-charged hole. Photosystem II has the job of filling the holes with electrons obtained by splitting water. Photosystem II seems to be composed of about 200 molecules of chlorophyll a, somewhat less than 200 of chlorophyll b, plus one reaction centre molecule of chlorophyll a designated P680 because of the wavelength at which it absorbs light. Apart from these differences, the organization is probably quite similar to that of photosystem I. Light energy collected is channelled to P680, which passes an electron to an acceptor, this time called 'Q', again of uncertain chemical nature. Substance Q can pass electrons to a chain of electron carriers—substances that can exist in either oxidized or reduced forms—and eventually to the chlorophyll in photosystem I. In this way, electrons originally in molecules of water and having a comparatively low energy, are given two 'lots' of energy to raise them to the level at which they can reduce $NADP^+$. We can summarize the sequence of electron movements in green plant photosynthesis as follows:

$$\text{water} \rightarrow \text{photosystem II} \rightarrow \text{'Q'} \rightarrow \begin{array}{c}\text{sequence of}\\ \text{electron carriers}\end{array}$$

$$\rightarrow \text{photosystem I} \rightarrow \text{'X'} \rightarrow \begin{array}{c}\text{sequence of}\\ \text{electron carriers}\end{array} \rightarrow NADP^+$$

The next problem is to see how ATP is generated from ADP and inorganic phosphate during photosynthesis.

How is ATP synthesized during photosynthesis?

The only possible source of the energy required to join ADP and inorganic phosphate together during photosynthesis is sunlight. We know that light energy can be converted to potential energy in electrons: if such high-energy electrons are allowed to 'flow' downhill to some electron acceptor, this energy will be released. How can this energy be used to bring together ADP and inorganic phosphate? We only understand the process in outline, but what we do know about it makes us reasonably sure that it is a very similar process to that going on when *mitochondria* produce ATP during the oxidation of foodstuffs. Only the barest outline will be given here therefore and we shall deal with the matter in much greater detail when we come to deal with mitochondria.

There are present in chloroplasts, members of a class of iron-porphyrin proteins called *cytochromes*. In these proteins the porphyrin prosthetic group is attached covalently to the protein and typically the protein is closely associated with membrane structures. The cytochromes are pinkish in colour, and because of their iron atom, they can act as electron carriers, existing in either the oxidized (Fe^{3+}) or the reduced (Fe^{2+}) states. Cytochromes are also found in mitochondria.

During the course of photosynthesis electrons can flow 'downhill', i.e. down a gradient of potential whereby they pass through a series of carriers of lower and lower potential. During this downhill passage energy is released. In both chloroplasts and mitochondria the energy released is captured in a gradient of protons and electrical potential across the membrane. As we shall see later (p. 118) dissipation of this gradient leads to the synthesis of ATP.

Although such ideas are difficult to visualize and understand, there is now much experimental evidence to indicate that this is the way in which ATP is generated in chloroplasts and mitochondria (Box 3.5). For example, if the membrane, mitochondrial or photosynthetic, is disrupted, then no ATP synthesis takes place. On the other hand, if isolated, but intact, chloroplasts are subjected to a sudden change in pH, effectively generating a gradient of protons across the membrane, the ATP is formed *in the dark*. Figure 3.16 shows an overall scheme for photosynthesis.

Box 3.5 *The purple membrane of Halobacteria*

Halobacteria are found in regions where there is a very high salt concentration such as in salt lakes and in pans where sea water is evaporated to produce salt. These bacteria actually require high salt environments for growth and live successfully in 4.3 M NaCl (compare ordinary sea water which is about 0.6 M). Part of the cell membrane of these bacteria contains a purple membrane protein called bacteriorhodopsin which is associated with lipids. This protein contains material very closely similar to the visual protein, rhodopsin, found in the sensitive cells of vertebrate eyes.

In the presence of oxygen, Halobacteria make ATP by oxidative phosphorylation (p. 114) and have an ATPase complex orientated so that it 'points' into the cell. However, if the oxygen supply becomes depleted, the bacteria may switch to a photosynthetic way of making ATP—but a very simple one. They have no chlorophyll, but instead the action of light on the purple membrane causes it to act as a proton pump. Protons are pumped from the inside to the outside of the cell, light energy being used to make the proton gradient. The protons then flow back via the specially orientated ATPase generating ATP inside the cell. This is probably the simplest chemiosmotic system known.

Cyclic photophosphorylation

The type of electron flow described above is 'non-cyclic', i.e. electrons from water are passed eventually to NADPH and some ATP is also generated in the process. This is *non-cyclic photophosphorylation*. We can, however, envisage a way in which electrons could take a cyclic path, generating ATP but not NADPH. If light energy raised an electron to a higher energy level in 'X', and it then dropped back through a system of electron carriers eventually to refill the positively-charged hole in chlorophyll, energy would be released in the downhill flow through the carriers and ATP could be

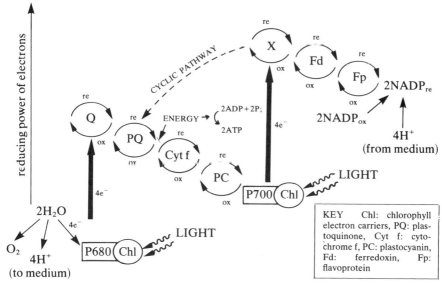

Figure 3.16 Scheme to show overall reactions during the light phase of green plant photosynthesis. (Redrawn from Keeton's 'Biological Science', W. W. Norton)

generated as with the non-cyclic system. Such a flow of electrons coupled to ATP production is called *cyclic photophosphorylation*, but its quantitative significance in chloroplasts is difficult to assess.

Step-by-step reactions

You will have noticed that in the schemes described above, reactions and electron flow go on step-by-step. This is typical of biochemical systems and applies particularly when energy is being released. Instead of energy being released in one step—effectively the same as an uncontrolled explosion—it is released in small, manageable packages, in controlled amounts.

In all organisms other than photosynthetic ones, the major method of ATP production is to pass high-energy electrons down a gradient of potential. In the hours of darkness photosynthetic organisms have to use this method of ATP production too. The high-energy electrons are taken from food materials or from the stores (for example starch in plants and glycogen in animals) and dropped to a lower level through carrier systems. Eventually the electrons combine with oxygen and protons to form water. This completes the cycle started when water was split to provide electrons in non-cyclic photophosphorylation.

3.4. ATP production in heterotrophs

Heterotrophic organisms obtain their energy by oxidizing foodstuffs in reactions some of which are coupled with the formation of ATP from ADP and inorganic phosphate. Although we, as heterotrophs, can eat a great variety of different foods, in the body these are broken down very efficiently into

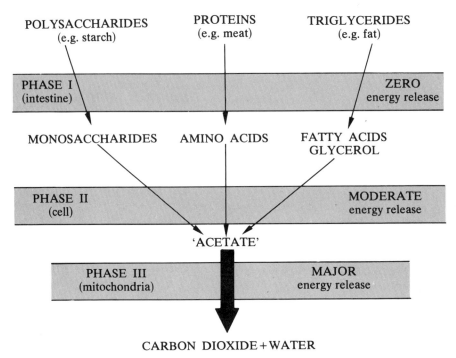

Figure 3.17 The 'metabolic funnel' for the catabolism of food materials. Each of the arrows represents a number of individual, enzyme-catalyzed reactions. It would be possible to construct an almost identical diagram for the utilization of *stored* materials in both animals and plants

Phase I takes place in the intestine and the 'monomer units' released by digestion pass via the blood to the liver and other tissues. No usable energy is released in this phase.

Phase II takes place in the cell, and partially in the mitochondria. 'Acetate' means a derivative of acetate, acetyl coenzyme A. A moderate amount of energy (as ATP) is obtained during this phase.

Phase III takes place entirely in the mitochondria and is common to all types of food material. The majority of the energy available from the oxidation of food and storage materials is released (as ATP) in this phase

their component parts and then channelled into a single metabolic pathway common to all foodstuffs, during which oxidation takes place. The overall process is referred to as the 'metabolic funnel' and is shown in Figure 3.17. A similar diagram could be drawn for the utilization of stored materials such as starch or glycogen in both animals and plants.

In phase I the material, mostly macromolecules, is broken down to the monomer units by digestive enzymes including those of the stomach and small intestine (Figure 3.18). In phase II these monomer units are split to produce a derivative of acetic acid, with some production of ATP. The major oxidations take place in phase III and it is in this common terminal pathway, occurring in the mitochondria, that the majority of the ATP production takes place. For this reason the mitochondria have been called the 'powerhouses' of the

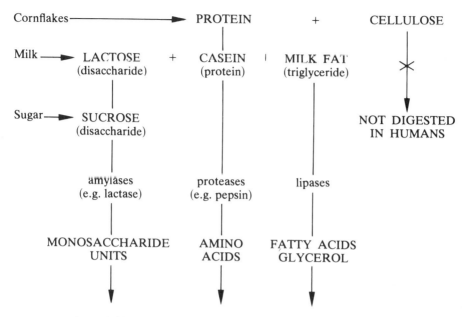

Figure 3.18 How your breakfast is fed into the metabolic funnel

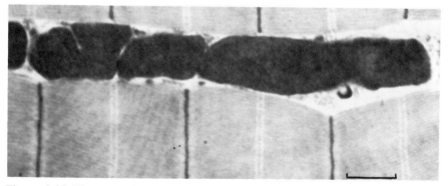

Figure 3.19 Electron micrograph of section through a piece of Tsetse fly muscle showing mitochondria packed into the spaces between the muscle fibres. Length of line 1 μm. (*Photograph kindly supplied by Dr M. Anderson, Department of Zoology, University of Birmingham*)

cell. The more active a cell the more mitochondria it contains. In insect flight muscle the mitochondria are tightly packed alongside the contractile fibres (Figure 3.19).

We are going to start by looking at the common steps first, i.e. phase III; and only then go back to see how all other pathways feed into this. This will give a sense of overall direction in metabolism: looking at the catabolism of any given food or storage material we shall know that ultimately it feeds into phase III of metabolism.

How is ATP produced during phase III?

The first major way of making ATP we saw was photosynthesis: the second major way is the process taking place in mitochondria. The first process, is called *photophosphorylation*: this second process is called *oxidative phosphorylation*. Photophosphorylation is, of course, only available as a way of making ATP in photosynthetic organisms and so oxidative phosphorylation is of major importance to all other organisms, as well as to plants in the dark!

The overall process of oxidative phosphorylation is the sum of two coupled processes: (1) *oxidation* of certain compounds derived from food or storage materials, and (2) *phosphorylation* of ADP by inorganic phosphate to give ATP. It would be satisfying to know exactly *how* these two process are coupled. However, despite a great deal of intensive research over the last 30 years, biochemists still only have a partial answer. We will look first at what is known of these two processes separately and then try to see how they are coupled.

Oxidation in mitochondria

Food and storage materials represent forms of carbon which are highly reduced compared with carbon dioxide, the end product of their metabolic breakdown (compare (CH_2O) with CO_2). Somewhat surprisingly, however, 'oxidations' in metabolism rarely involve oxygen itself. More often than not enzyme-catalyzed reactions occur which are in reality 'dehydrogenations', that is, reactions in which hydrogen atoms and electrons are removed from a compound rather than ones in which oxygen is added. The compound therefore becomes 'less reduced' which is equivalent to being 'more oxidized'. The enzymes involved are called 'dehydrogenases'.

Oxygen, as we well know, *is* absolutely essential for the life of most organisms, although there are bacteria which can exist anaerobically. However, it only participates in metabolic breakdown reactions at the very last stage. This is when the hydrogen atoms and electrons, mentioned above, unite with oxygen to form water, which we can look upon as the 'other' product of metabolic breakdown besides carbon dioxide (Figure 3.20).

Let us look at this process of 'oxidation' in a little more detail. The electrons removed from food and storage materials have a high potential energy, that is, they have the potential to combine with protons and oxygen to form water. When this happens energy is released in the same way that petrol (another highly-reduced form of carbon) can be oxidized with the release of energy

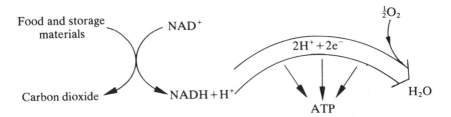

Figure 3.20 The major energy-yielding steps during the catabolism of food and storage materials involve dehydrogenations. Oxygen itself only participates at the very last step

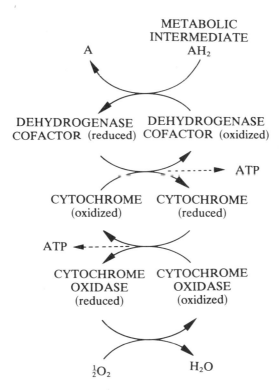

METABOLIC
INTERMEDIATE

A   AH$_2$

DEHYDROGENASE   DEHYDROGENASE
COFACTOR (reduced)   COFACTOR (oxidized)

ATP

CYTOCHROME   CYTOCHROME
(oxidized)   (reduced)

ATP

CYTOCHROME   CYTOCHROME
OXIDASE   OXIDASE
(reduced)   (oxidized)

$\frac{1}{2}O_2$ H$_2$O

Figure 3.21 Simplified scheme showing route taken by electrons. At each step the reduced form of one carrier reduces (i.e. passes electrons to) the oxidized form of the next. There are many more such transfers than shown here

(as heat in this case). In living systems, the electrons are dropped little by little from their high potential level to that of oxygen, and this is achieved by having a chain of carrier molecules, including the cytochromes mentioned previously. At each step some energy is released. Energy released in this way is 'trapped' in reactions which result in the formation of ATP from ADP and inorganic phosphate (Figure 3.21). The chain of compounds through which the electrons pass is called the *electron transport chain*, and is simply a group of compounds each capable of existing in oxidized or reduced states.

Many years ago it was noticed that the absorption spectrum of mitochondria changed when the medium in which they were suspended was deoxygenated. Later, chemical analysis allowed the identification of the compounds that carried electrons and finally passed them to oxygen. In the absence of oxygen, of course, all the carriers become reduced, i.e. have electrons ready to pass on. We know now that there are several types of compound in the mitochondria which carry electrons (or electrons *and* protons in some cases): NAD$^+$ (which is identical to NADP$^+$ except that it lacks one phosphate group), the flavins, the quinones, the iron–sulphur proteins and the cytochromes (Table 3.1). Probably these are arranged, along with the appropriate enzymes, in a special orientation in the inner membrane of the mitochondria.

Table 3.1 The hydrogen and electron carriers of the electron transport chain

Name	Reaction	Outline structure
Nicotinamide adenine dinucleotide	$NAD^+ \rightleftharpoons NADH + H^+$	adenine nicotinamide \ / ribose—℗—℗—ribose (compare with $NADP^+$, Fig. 3.10)
Flavins (e.g. flavin adenine dinucleotide (FAD) and flavin mononucleotide (FMN))	$FAD \rightleftharpoons FADH_2$	adenine flavin \ / ribose—℗—℗—ribitol (FAD)
Iron–sulphur proteins	(Fe.S) complex carries electrons	protein S S S Fe Fe S S S protein
Coenzyme Q (ubiquinone)	$CoQ \rightleftharpoons CoQH_2$	(long hydro-phobic side chain)
Cytochromes	$Fe^{3+} \rightleftharpoons Fe^{2+}$	—N N— Fe —N N— (in porphyrin ring covalently linked to protein)

The electron carriers of mitochondria

The electron-carrying system of the mitochondria can be thought of as a series of dehydrogenase enzymes (with one *oxidase* enzyme at the end of the chain). Each of these enzymes has a cofactor that can exist in the oxidized or the reduced state. The first member of this chain accepts protons and electrons from metabolic intermediates during the catabolism of food and storage molecules: the last member of the chain passes electrons on to oxygen. We can trace the path of an electron through such a series of carrier molecules (Figure 3.22).

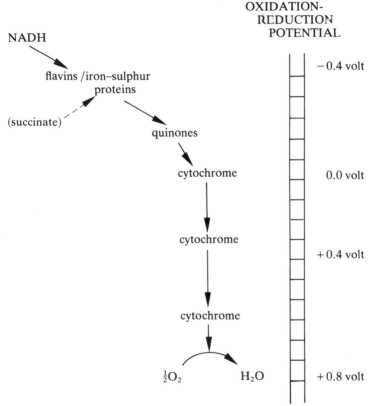

Figure 3.22 The passage of electrons through the electron transport chain involves the passing of electrons *down* a potential gradient. Hydrogen itself (a good reducing agent) would appear at the top of this scheme with a high, negative oxidation–reduction potential. Note that this scheme is simply another way of presenting the information in Fig. 3.21

Many metabolites are capable of feeding electrons and protons to NAD^+, but others pass electrons and protons to the flavoproteins instead. Both of these steps therefore are linked directly to compounds undergoing metabolic breakdown. The flavins, or flavoproteins, are yellow-coloured compounds. They are closely associated with the iron–sulphur proteins of the mitochondria whose iron atoms carry electrons. Eventually, electrons are passed on to a quinone called *ubiquinone* which can donate its electrons to cytochromes.

There are several types of cytochrome in mitochondria which differ in properties: they are designated by letters, thus cytochrome *a*, *b* and *c*. In each an iron atom is held in a porphyrin ring system and is attached to a protein molecule. The cytochromes form a series of increasingly powerful oxidizing agents—that is compounds that will accept electrons. Electrons appear to be passed in the sequence:

$$\text{cytochrome } b \rightarrow \text{cytochrome } c \rightarrow \text{cytochrome } a.$$

(Unfortunately the cytochromes were named according to their spectra and their spectra do not happen to coincide with the order in which they function in the mitochondria!)

The very last step in the sequence, the passing of electrons to oxygen, is catalyzed by an enzyme called *cytochrome c oxidase*. This is a very complex unit consisting of proteins, cytochromes *a* and a_3, as well as copper.

How is ATP produced?

While all this is going on, ATP is formed from ADP and inorganic phosphate. Although it is believed that the processes involved in the formation of ATP in the mitochondria and in the chloroplasts are closely similar, neither process is completely understood. It *is* known that for every pair of electrons taken from a metabolite and fed through NAD^+ into the electron transport chain in mitochondria, 3 molecules of ATP are synthesized from ADP and inorganic phosphate. Some metabolites, however, cannot supply electrons at a sufficiently high potential to reduce NAD^+, and reduce a flavoprotein instead, bypassing one step in the chain. When this happens only 2 molecules of ATP are generated.

There have been many theories proposed to try to describe how electron transport is *coupled* to ATP formation. The problem is that the reaction:

$$ATP + H_2O \rightleftharpoons ADP + P_i$$

is well-nigh irreversible at neutral pH in free solution. The equilibrium constant is very high, and we saw that a reason for this was that both ADP and inorganic phosphate carry negative charges and so tend to repel each other. The reaction does not, in fact, go on in free solution in living organisms, but occurs in association with solid-state devices, the inner mitochondrial membrane or the chloroplast membrane.

Mitchell's theory

Many people now accept that ATP production results from the utilization of energy held in a *gradient of charge and concentration*. Ideas along these lines were originally proposed by the British biochemist, Dr Peter Mitchell, some years ago. He was awarded the Nobel Prize in 1979 in recognition of his contribution to science. The ideas themselves are fairly simple but understanding the detailed chemical mechanism is impossible at present because we have too little information about the structures of the membranes involved, those of the mitochondria and chloroplasts.

Mitchell's theory is called the *Chemiosmotic Theory*, and is based upon the very simple idea that to form and maintain a concentration gradient requires energy. If the gradient is allowed to dissipate the energy will be released. It is proposed that the electron and proton carriers described above are arranged in a highly-ordered way within a membrane. Because of this high degree of organization at the molecular level, protons can be translocated to one side of the membrane when electrons flow through the carriers. In the case of mitochondria, electron flow results in the expulsion of protons from the mitochondria. The medium outside thus becomes richer in protons (lower pH) as well as in positive charges, than the inside. Energy released in passing electrons down a *potential gradient* is now stored in a *concentration gradient*. A key factor in this is, of course, that the membrane must be *impermeable* to protons. Mitchell proposed that there were certain sites in the membrane at which enzymes called 'ATPases' were found. An ATPase catalyzes the hydrolysis of ATP to ADP and inorganic phosphate—exactly the opposite of what is needed if ATP is to be *made*, you might think. But enzymes catalyze

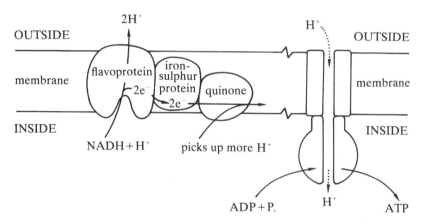

Figure 3.23 The chemiosmotic theory. The diagram shows a possible but no doubt highly-simplified way in which oxidative phosphorylation might take place in mitochondria. The carriers of the electron transport chain are arranged in the inner mitochondrial membrane. Some of these traverse the whole membrane, others tend to sit on one side, while yet others may be completely embedded within the membrane. Exactly where they sit depends on the degree of hydrophobicity of the different regions on their surfaces.

Two stages in the overall process can be recognized: (1) *A gradient of protons is generated across the membrane.* Both protons and electrons are fed in from $NADH + H^+$, but some of the carriers *only* carry electrons. Because of the physical arrangement of the carriers in the membrane, protons can be expelled to the outside while electrons travel on within the membrane. As they pass through quinones situated near the inside of the membrane, more protons may be picked up and so on. (2) *Passage of protons through channels in ATPase enzyme molecules in the membrane, from outside to inside, causing ATP to be formed from ADP and P_i.* The energy to drive this process is the energy that was previously stored in the gradient of protons, and of charge, across the membrane. How ATP synthesis takes place on the surface of this enzyme is uncertain: the protons passing close to the active site may cause a conformational change.

A similar scheme can be drawn for the production of ATP in chloroplasts during photosynthesis

reactions in both directions—that is they catalyze the attainment of equilibrium. Mitchell suggested that there were channels in the membrane through which protons could pass. When such an event occurred the membrane ATPase could 'work backwards' favouring the *formation* of ATP. Possibly the protons could affect the *local* environment within the membrane near the ATPase, perhaps causing a conformational change (Figure 3.23).

For any scientific theory to be acceptable as an explanation of the observed facts, it must be testable experimentally. Can this be done for Mitchell's theory? Obviously, if the membrane were to be disrupted or 'punctured', then the proton gradient would be lost. If the theory is correct, no ATP should be made. This is found: disrupted mitochondria will not make ATP, and certain chemicals, when added to intact mitochondria, make the membrane 'leaky' to protons, and also stop ATP synthesis. Many such chemicals are known, and are called 'uncoupling agents' (Box 3.6). It would also be desirable to detect experimentally a difference in pH on either side of the membrane. Experiments have been done which seem to confirm this, but technically they are difficult to do. Many experiments therefore support the chemiosmotic theory, despite initial scepticism by many eminent biochemists. Before we can describe the intricate workings of ATP production in these organelles we must have a very much more detailed knowledge of the structure of mitochondrial and chloroplast membranes.

Box 3.6 Uncoupling agents and brown fat

Many known chemicals will 'uncouple' oxidation from phosphorylation. In this situation the energy released by oxidation of food materials is converted into heat instead of being used to manufacture ATP. The first uncoupling agents were discovered when it was noticed that women working in munitions factories during the First World War became very thin regardless of how much they ate. It was found that they were inhaling certain nitro-compounds from the explosives in the course of their work.

Dinitrophenol is such an uncoupling agent, but many others are known. Dinitrophenol was used for a time as a slimming agent. It works quite well but unfortunately has very severe side-effects, and its use is now prohibited.

$$O_2N-\!\!\left\langle\bigcirc\right\rangle\!\!-OH$$
$$NO_2$$

Warm-blooded animals maintain a constant body temperature. The heat evolved as a by-product of normal metabolism is usually adequate for this purpose and animals have mechanisms for regulating how much heat they lose to the environment. However, warm-blooded animals have a tissue called 'brown fat' which differs from the ordinary 'storage fat' in having many more mitochondria in the cells and a much richer blood supply—hence its reddish-brown colour. Brown fat is especially rich in hibernating animals and neonates of animal species born without fur. It is believed that certain naturally-occurring compounds can act as uncoupling agents in brown fat with the result that heat is released rather than ATP generated. Brown fat is therefore a heat-producing tissue in situations when it is necessary to raise the body temperature—such as when an animal is rousing from a period of hibernation. It could be that 'fat' humans have less brown fat than thin ones!

Fuelling the powerhouses

It is obvious that, given a supply of the correct oxidizable substrates, the process of ATP production by mitochondria will keep going. Much of the catabolic breakdown of food and storage materials (the metabolic funnel) is geared up to *getting these materials ready* so that they can take part in reactions in which NADH or reduced flavins are produced. So our next question is: how are the reduced cofactors produced as a result of the breakdown of foodstuffs and storage materials?

The metabolic product arising from the breakdown of almost all of these materials is a derivative of acetic acid, *acetyl coenzyme A*. The coenzyme A part of this molecule is a complicated organic compound; we shall simply think of it as a 'handle' by which enzymes can recognize it. The acetyl part, CH_3CO-, represents a two-carbon (C_2) fragment, derived from foodstuffs and storage materials (Figure 3.24). In the section that follows it will be seen how this C_2 unit is converted to 2 molecules of carbon dioxide while simultaneously, reduced coenzymes are generated. The sequence of reactions by which this is achieved is perhaps the most famous sequence of metabolic reactions: the Krebs Cycle, also called the Citric Acid Cycle, alias the Tricarboxylic Acid Cycle (TCA cycle).

Coenzyme A is made up of the nitrogen-containing base, adenine, joined to ribose joined to two phosphate groups joined in turn to pantothenic acid (a vitamin). The pantothenic acid unit carries a mercaptoethylamine unit, and it is to the sulphydryl on this that a acetyl unit can be attached. An outline structure is shown below:

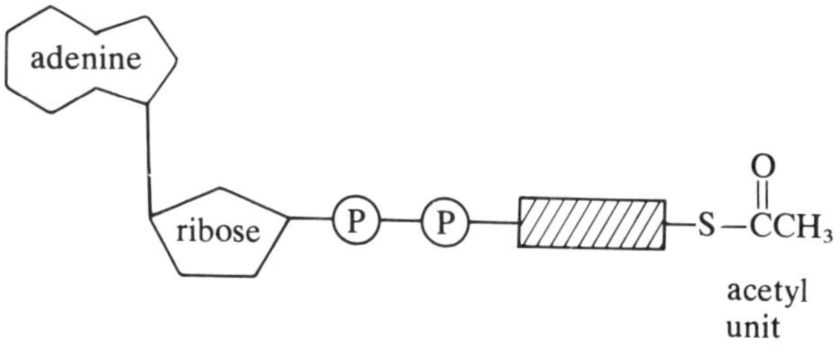

This coenzyme can be thought of in a similar way to $NADP^+$ (Fig. 3.10), that is as having a 'handle' recognizable by enzymes, and a 'business end', in this case a $-SH$ group capable of combining with an acetyl group. Energy in the form of ATP has to be 'spent' when making acetyl coenzyme A from acetate and coenzyme A. The acetyl unit is therefore somewhat unstable and is readily removed by hydrolysis. Because of this high reactivity, it is said to be 'activated'. Acetyl coenzyme A used to be called 'active acetate'

Figure 3.24 Structure of acetyl coenzyme A

3.5. The tricarboxylic acid cycle

A cycle of enzyme-catalyzed reactions was discovered in 1937 by Sir Hans Krebs (Figure 3.25). This cycle effectively oxidizes a C_2 unit completely to carbon dioxide. This is the last stage in *respiration*, the outward signs of which are the consumption of oxygen and the liberation of carbon dioxide.

Figure 3.26 is a very much simplified way of looking at the TCA cycle and shows how carbon atoms are 'juggled with' in order to achieve certain aims.

Figure 3.25 The late Sir Hans Krebs with a borrowed cycle

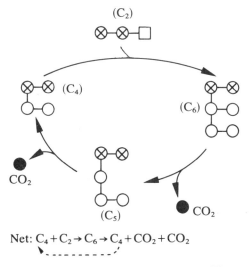

Net: $C_4 + C_2 \rightarrow C_6 \rightarrow C_4 + CO_2 + CO_2$

Figure 3.26 The tricarboxylic acid cycle—overall strategy. The net effect of the cycle is to convert one molecule of acetate (as acetyl coenzyme A) into 2 molecules of carbon dioxide. The C_4 intermediate is regenerated

One aim is to convert an acetyl coenzyme A C_2 unit to 2 molecules of carbon dioxide. The more profound aim, however, is to produce substances that can pass protons and electrons on to coenzymes such as NAD^+ and flavins, and subsequently to the electron transport chain. Finally, since the last intermediate of the pathway of the reactions is *also* the first intermediate in the pathway, the pathway is cyclic. A very small amount of the C_4 compound (Figure 3.26) will keep the cycle going because it is regenerated.

Figure 3.27 shows a more detailed picture of the reactions of the TCA cycle. At four points in the cycle reduced coenzymes are generated. Each of these can enter the electron transport chain generating 3 ATP molecules in the case of NADH and 2 in the case of reduced flavins. One 'turn' of the cycle therefore generates at least 11 molecules of ATP. All of the reactions

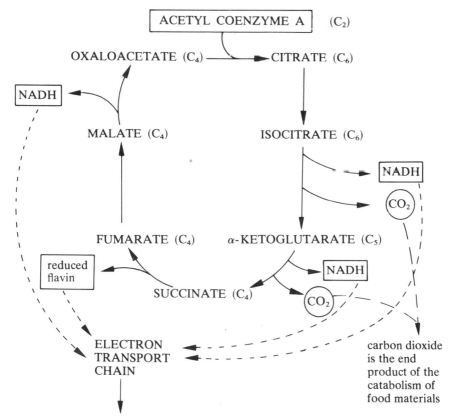

Figure 3.27 The tricarboxylic acid cycle shown in more detail with emphasis on the steps in which reduced coenzyme is produced

of the TCA cycle go on inside the mitochondria, and so there is no difficulty in the coenzymes passing on protons and electrons.

You may wonder how the acetyl unit, CH_3CO-, can produce *four* pairs of protons and electrons. The answer is that in addition to protons and electrons being stripped off from intermediates, another little metabolic trick is being used. This trick is to add water to a compound (i.e. hydrate it) in one step, and then, in the next step, remove protons and electrons, *leaving the oxygen behind* (Figure 3.28). This is another way of oxidizing a compound without using molecular oxygen: it occurs elsewhere in metabolism too. Also, at one step of the tricarboxylic acid cycle the equivalent of one ATP is produced by a 'substrate level phosphorylation' (see p. 130): therefore the net yield of 'high-energy bonds' from acetyl coenzyme A passing round the cycle is actually 12 not 11.

One of the main experimental findings that led Krebs to propose a cycle in 1937 was that respiration was inhibited by malonate. As it happens malonate is closely similar to succinate, one of the intermediates of the TCA cycle. Malonate acts as a competitive inhibitor to the enzyme succinate dehydrogenase, stopping the cycle (Figure 3.29). This shows how competitive inhibitors can be of use to biochemists in discovering the secrets of metabolism.

$$
\begin{array}{l}
CO_2^- \\
| \\
CH_2 \\
| \\
CH_2 \\
| \\
CO_2^-
\end{array}
\xrightarrow[\substack{\text{succinate} \\ \text{dehydrogenase}}]{-2H^+, -2e-}
\begin{array}{l}
CO_2^- \\
\diagup \\
CH \\
\parallel \\
CH \\
\diagdown \\
CO_2^-
\end{array}
\xrightarrow[\text{fumarase}]{+H_2O}
\begin{array}{l}
CO_2^- \\
| \\
CHOH \\
| \\
CH_2 \\
| \\
CO_2^-
\end{array}
\xrightarrow[\substack{\text{malate} \\ \text{dehydrogenase}}]{-2H^+, -2e-}
\begin{array}{l}
CO_2^- \\
| \\
C=O \\
| \\
CH_2 \\
| \\
CO_2^-
\end{array}
$$

succinate   fumarate   malate   oxaloacetate

Figure 3.28 Three steps of the tricarboxylic acid cycle in detail

Malate,
$\begin{array}{l} CO_2^- \\ \diagup \\ CH_2 \\ \diagdown \\ CO_2^- \end{array}$
, is closely similar in shape to succinate,
$\begin{array}{l} CO_2^- \\ \diagup \\ CH_2 \\ | \\ CH_2 \\ \diagdown \\ CO_2^- \end{array}$

and inhibits the enzyme succinate dehydrogenase competitively. Malonate is a meta-bolic poison. The fact that it almost completely blocks respiration, that is oxidation of materials through the TCA cycle, suggests that the TCA cycle is the *only* major pathway of respiration. Krebs offered the following 'proof' that the TCA cycle *was* a cycle and not just a collection of unrelated reactions occurring in cells. We start from the knowledge that the succinate dehydrogenase reaction is reversible. Adding fumar-ate to minced tissue is therefore bound to lead to the production of succinate *either* (1) by a reversal of the succinate dehydrogenase reaction, *or* (2) by fumarate going round the cycle. In the presence of malonate, fumarate is *still* converted to succinate. Since malonate blocks the succinate dehydrogenase reaction (whichever way it is operating), only the second way of forming succinate is possible now—that is round the cycle. Therefore the cycle must exist!

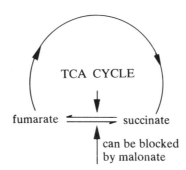

TCA CYCLE

fumarate ⇌ succinate

can be blocked
by malonate

Figure 3.29 Malonate inhibits the tricarboxylic acid cycle

It also shows incidentally that the TCA cycle is the *only* major terminal catabolic pathway in almost all organisms. That is, since respiration is virtually stopped by malonate, there are no alternative routes. Other inhibitors are also known (Box 3.7).

Now we go back a step further, to find out how acetyl coenzyme A is produced from food and storage materials.

Box 3.7 Fluoroacetate, a potent poison

Malonate is a metabolic poison because it interferes with the succinate dehydrogenase reaction. This observation showed that the TCA cycle was the *only* major terminal pathway of metabolism. Another compound that inhibits the TCA cycle is fluoroacetate.

$F{-}CH_2{-}CO_2^-$	$H{-}CH_2{-}CO_2^-$
fluoroacetate	acetate

When fluoroacetate is fed into the TCA cycle, fluorocitrate is formed, instead of citrate. However, the enzyme that normally deals with citrate will not deal with fluorocitrate and the cycle soon grinds to a halt.

Recently, factories producing fluorinated compounds have allowed fluoroacetate-containing effluents to escape and contaminate near-by rivers and pastures. Cattle have died as a result of this. It has been found that the lethal dose of fluoroacetate for a rat is about 0.2 mg for every kg of body weight. Strangely the leaves of certain plants found in Africa, Australia and South America actually contain fluoroacetate. Needless to say such leaves are highly poisonous to animals.

3.6. Converting fat to acetyl coenzyme A

Fat in the diet, as well as stored fat, is neutral fat or triglycerides (see p. 49). Foods rich in fat include meat, butter and milk as well as plant products such as olive oil. Triglycerides are hydrolyzed by a lipase enzyme in the small intestine to give fatty acids and glycerol. A similar process occurs in adipose tissue to *mobilize* fat when this is needed (Figure 3.30). The glycerol *is* used in metabolism, although we shall not consider it further here, concentrating instead on the long-chain fatty acids produced. These are brought, either from the intestinal epithelium or from the adipose tissues, via the blood, to the liver and other tissues, where catabolism takes place in the mitochondria.

Breakdown of fatty acids takes place in three stages:

(1) Activation.
(2) Breakdown to acetyl coenzyme A.
(3) Catabolism of the acetyl unit by the TCA cycle.

We already know about the third stage.

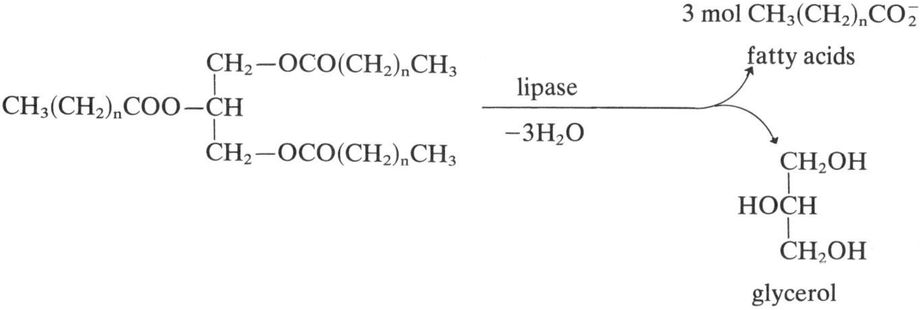

Figure 3.30 Release of fatty acids from triglycerides. The action of lipases on neutral fat (triglycerides), either in food or storage materials releases fatty acids which can be oxidized to provide energy

In order to be broken down, fatty acids have first to be activated. This simply means that they have to have a molecule of coenzyme A added to their carboxyl groups to form 'acyl coenzyme A' derivatives. It 'costs' the cell ATP to drive this activation step, but the investment is well worthwhile as will be seen shortly. As well as activating the fatty acids, making them ready for breakdown, adding coenzyme A may also make the fatty acids 'safe'. Free fatty acids are, of course, soaps or detergents that can disorganize membranes, and fatty acid breakdown goes on in membrane-bounded structures, the mitochondria.

Activated fatty acids are catabolized by being 'nibbled away' in C_2 units. Their breakdown sequence does not form a cycle, but rather a *spiral*. Only about four enzymes are used, but these act in a short sequence, over and over again, and each time a C_2 unit is removed from the long-chain fatty acid. The C_2 units are acetyl coenzyme A molecules. During this sequence of reactions, NADH and reduced flavins are produced, which give rise to ATP by passing their protons and electrons to the electron transport chain. This sequence of reactions, you may be surprised to know, is already familiar to you. It is exactly the same as the one that occurs during some of the steps of the TCA cycle. Firstly, protons and electrons are removed; then water is added; then another pair of protons and electrons is removed leaving the oxygen behind (Figure 3.31).

The fatty acids occurring in beef fat mostly have 16 or 18 carbon atoms in the chain. Hence, each can produce eight or nine C_2 units, as a result of the breakdown steps as described above. Each of the breakdown steps can produce

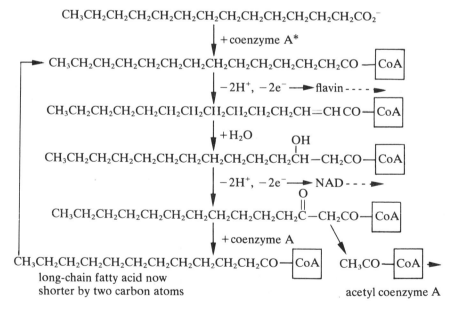

* Note: the addition of coenzyme A to a fatty acid or to acetate is an energy-requiring reaction and a molecule of ATP has to be hydrolyzed to drive the reaction

Figure 3.31 Catabolism of long-chain fatty acids—the fatty acid spiral

one NADH and one reduced flavin, equivalent to 5 ATP molecules. Further-more, each of the acetyl coenzyme A molecules produced can feed into the TCA cycle, producing 12 ATP molecules. So you can see that one molecule of C_{18} fatty acid produces $(8 \times 5) + (9 \times 12)$ ATP molecules, less one or two molecules spent in the original activation with coenzyme A.

Most naturally-occurring fatty acids have an even number of carbon atoms, but a few have an odd number. There is a special mechanism, involving a couple of enzymes, for dealing with them. Most naturally-occurring fatty acids have one or more double bonds. You will see from Figure 3.31, that they can take part in the scheme of breakdown by 'skipping' the first step. The whole process is therefore very efficient: only a dozen or so enzymes are needed to extract the energy from several hundred different types of fat in the diet. We may also note at this stage, that the catabolism of fatty acids can only take place aerobically, that is, in the presence of oxygen. In the absence of oxygen, the supply of coenzymes (NAD^+, flavins) soon runs out because their reduced forms cannot be reoxidized.

Now we turn to the carbohydrates as suppliers of acetyl coenzyme A. Good though fat is as an energy source, there are reasons why carbohydrates often have to be used instead.

3.7. Converting carbohydrate to acetyl coenzyme A

Carbohydrates are the major energy-supplying nutrients for the majority of the world's population. Starch in cereals and other crops is much cheaper and more economical to produce than fat and protein in meat.

Digestion of starch in the gastrointestinal tract releases glucose, which passes through the intestinal epithelium into the blood and hence finds its way to the liver. Free glucose is equally acceptable and may be used to feed unconscious patients by means of an intravenous drip. Stores of carbohydrate such as glycogen can be mobilized to produce glucose too.

The process of breakdown of glucose was first studied in yeast cells and later in muscle tissue. One of the earliest observations was that breakdown could occur *anaerobically*, that is, in the absence of oxygen, with energy (i.e. ATP) production. This is in marked contrast to fat breakdown and there is a very important consequence. *Only* carbohydrate can be used as an energy supply in the absence of oxygen such as may occur with yeasts and other micro-organisms in certain environments. However, even parts of the human body become partially anaerobic from time to time. For example, when voluntary muscle is contracting vigorously, all the blood is squeezed out of the muscle bed. In order to carry on working, the muscle must obtain energy anaerobically now that blood is no longer bringing oxygen. It does so by breaking down glucose anaerobically. The whole process of breaking down carbohydrate was first called *anaerobic glycolysis* because of this (*glyco* = sugar, *lysis* = breaking down). In fact, the process goes on equally well in the presence of oxygen, and so is more often than not simply called 'glycolysis'.

Glycolysis

Glycolysis is quite a complicated sequence of metabolic reactions, about a dozen of them altogether. We shall start by trying to discover the overall strategy of the pathway before looking at a few of the steps in more detail. We can summarize by saying that glycolysis results in the breakdown of glucose, a C_6 compound, to smaller compounds. Eventually acetyl coenzyme A may be produced. Along the way some ATP is generated anaerobically. This is the third, and only other way of making ATP (the other two were photophosphorylation and oxidative phosphorylation). We may note that all of glycolysis goes on in the cytoplasm of the cell and does not seem to require membranous structures such as mitochondria. In the first stage of glycolysis, glucose is activated by combination with phosphate and is then split to give two C_3 compounds. In the second stage the C_3 compounds undergo a number of transformations as a result of which ATP is generated.

Stage 1 of glycolysis

'Activation' of the glucose molecule means adding phosphate groupings, first one, then two, to give various 'sugar–phosphate esters'. It costs the cell ATP to do this, but as was the case with fatty acids, the investment is repaid later on. The C_6 molecule bearing two phosphate groups is now weakened sufficiently to be split in the middle to give two C_3 phosphates. This process, like all the others, is catalyzed by a specific enzyme.

The C_3 phosphates so produced combine with inorganic phosphate to give C_3 diphosphate which passes on to stage 2 (see Figure 3.32 and below).

Stage 2 of glycolysis

Each of the C_3 diphosphates donates a phosphate group to ADP to give ATP: the electron transport chain is not involved in this ATP production. After a series of further 'jugglings' with the C_3 compounds, the second phosphate too is donated to ADP, leaving a C_3 compound called *pyruvate*, $CH_3COCO_2^-$. The fate of this compound depends upon whether or not oxygen is present. If oxygen is present, pyruvate can lose carbon dioxide with the formation of acetyl coenzyme A. In the absence of oxygen other routes have to be taken (see below). The overall reactions are summarized in Figure 3.32.

It takes 2 molecules of ATP to make the sugar diphosphates in stage 1. Since 2 ATP molecules are generated per C_3 unit in stage 2, a total of 4 ATP result per glucose. Therefore, the net gain of ATP molecules per glucose molecule split is 2. This is not very many and the process may not look very efficient compared with oxidative phosphorylation. However, very little of the energy originally in glucose has actually been released during glycolysis: most of it is still locked up in C_3 compounds such as pyruvate. Nevertheless, it is a fact that a lot of glucose must pass through this pathway in order to generate appreciable amounts of ATP anaerobically. Set against this must be the fact that this is practically the *only* way to obtain ATP anaerobically!

Figure 3.32 Outline of the process of glycolysis by which glucose is broken down and metabolic energy (ATP) is made available. The central part of the diagram shows the overall strategy of the pathway, and the chemical structures of some of the intermediates are shown on the right. All of the stages are catalyzed by specific enzymes found in the cytoplasm of cells. Broken arrows indicate that more than one enzyme-catalyzed step takes place at that point. What happens to the pyruvate produced depends upon whether or not oxygen is available and upon the type of organism

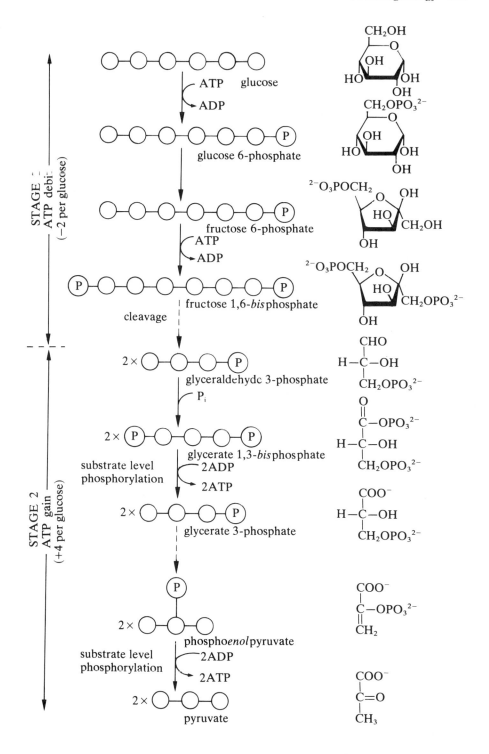

How is ATP generated during glycolysis?

We shall not look at all of the reactions of glycolysis in detail, but will concentrate on just two steps in order to see how ATP is generated. In Figure 3.33 are set out two of the reactions of glycolysis. The first compound, glyceraldehyde 3-phosphate is one of the C_3 compounds resulting from the splitting of the C_6 diphosphate in stage 1. As its name suggests, this C_3 compound contains an aldehyde grouping (—CHO). Aldehydes may be oxidized to carboxylic acids (—COOH) and energy is released in this oxidation. This is effectively what happens at this step in glycolysis, the energy being conserved as ATP. What *actually* happens is a little more complicated although it is not difficult to understand if we go step by step.

Firstly, in oxidizing aldehyde to acid, molecular oxygen cannot be used because, as we have already observed, the process of glycolysis is anaerobic. We now remind ourselves of the metabolic trick mentioned earlier—add water, then subtract hydrogen atoms leaving the oxygen behind. This would achieve the oxidation satisfactorily, but we need to go one more step to understand the subtleties of how ATP is produced. If, instead of adding water (H—OH), we added phosphoric acid (H—PO_4H_2), and then removed protons and electrons, instead of leaving behind oxygen, we would leave behind a phosphate group. What does this achieve? The phosphate that is left behind joined to the C_3 compound is not a phosphate ester but is an *acid anhydride* formed from two acids, phosphoric and a carboxylic acid (—COOH) (Figure 3.33). ATP is also an anhydride, but one formed from two phosphoric acid units. Such anhydrides are unstable, the C_3 phosphate very much so. In the presence of the appropriate enzyme it readily donates its phosphate group to ADP to produce ATP. The grouping 'left behind' on the C_3 compound is now a carboxylic acid instead of an aldehyde—so an oxidation has taken place

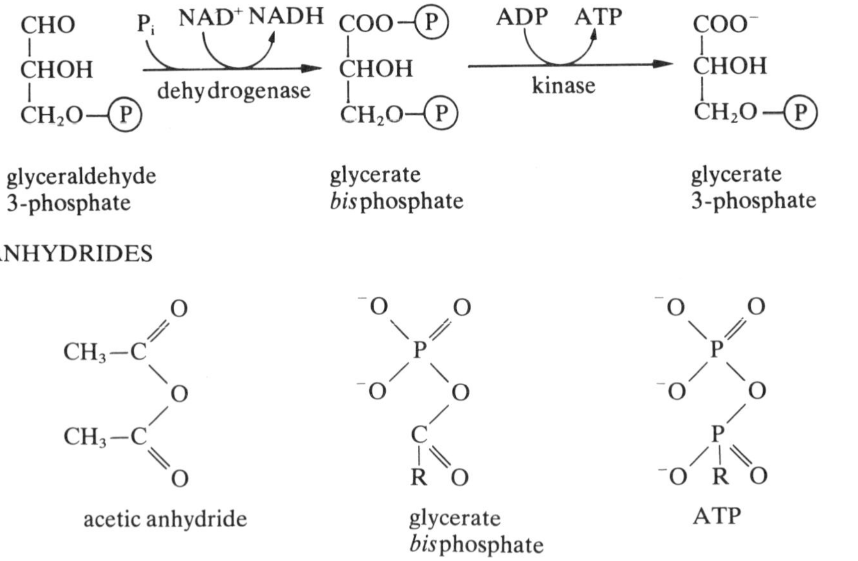

ANHYDRIDES

Figure 3.33 Two steps of the glycolysis pathway in detail showing how ATP is generated by substrate level phosphorylation

as we originally planned—and a molecule of ATP has been generated. In order to distinguish this type of ATP production from photophosphorylation and oxidative phosphorylation, it is called *substrate level phosphorylation* because a substrate molecule actually participates as a phosphate compound. Somewhere further on in the sequence of glycolysis another substrate level phosphorylation occurs.

Thus ATP can be generated in glycolysis; but we have a problem. Looking back at the sequence of reactions in Figure 3.33, the removal of protons and electrons is performed by the familiar coenzyme, NAD^+. There is very little NAD^+ in the cell, and once it has all been converted to NADH the reaction sequence shown in Figure 3.33 will come to a halt. In aerobic conditions there is no problem because the NADH can feed its protons and electrons into the electron transport chain, hence NAD^+ is regenerated. But this cannot happen under anaerobic conditions!

What happens to NADH in anaerobic conditions?

In anaerobic conditions glycolysis yields two compounds, pyruvate and NADH, and if the NADH is not reconverted to NAD^+, glycolysis stops. In vertebrate voluntary muscle, these two compounds get together, under the influence of an enzyme called lactate dehydrogenase, and pyruvate is reduced to lactate (Figure 3.34). The NADH is thereby reconverted to NAD^+, and lactate accumulates in the rapidly-contracting muscle. If vigorous contractions continue for a long time, much lactate accumulates and the muscle stops working—we call this 'cramp'. Normally before this happens the muscle either ceases to contract—because the animal stops running— or else the strength and rate of contractions drops. In either case blood can start to flow through the muscle bed, bringing oxygen and washing out lactate. Most of the lactate passes to the liver, where it is reconverted to pyruvate and NADH to be dealt with aerobically. During a period of vigorous exercise an *oxygen debt* arises (Box 3.8).

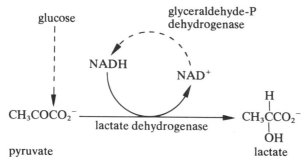

Figure 3.34 Lactate production in muscle under anaerobic conditions

Box 3.8 Oxygen debt, diving mammals and high altitudes

Oxygen debt. Tissues, especially muscle when contracting vigorously, often need so much energy so fast that the oxygen supplied by breathing is insufficient. After the period of violent muscular activity has ceased the 'oxygen debt' is gradually repaid. The lactate formed in muscle is transported to the liver. Some is oxidized, and some reconverted to glycogen for storage. Another part of the debt repayment consists of recharging the myoglobin of muscle with oxygen.

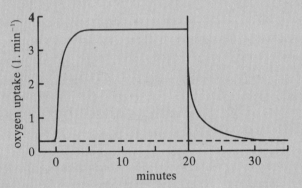

INITIAL PHASE (Oxygen debt). Slow increase due to delay in transport of blood from lungs to muscles and to delay in increasing cardiac output. Muscles receive insufficient O_2, obtain energy by anaerobic glycolysis: lactate formed.	SECOND PHASE (Steady state). O_2 uptake matches work output. Muscles working aerobically, or if not, lactate being washed out at same rate as formed. Blood lactate concentration elevated but constant.	THIRD PHASE (Recovery). O_2 uptake in excess of resting requirement: oxygen debt repaid. Myoglobin O_2 stores replenished and energy being used in reforming glycogen stores.

Types of muscle, red and white. Voluntary muscle can be either 'red' or 'white' or something in between. Chicken breast muscle is white, while chicken leg muscle is red. White muscle is constructed for short bursts of activity with strong contractions, sufficient to get a heavy chicken off the ground for a short flight. White muscle obtains energy by anaerobic glycolysis and the muscle mass has relatively few mitochondria and little myoglobin. Red muscle is constructed for repeated contractions for extended periods, e.g. in chicken leg muscle for the chicken to run around all day searching for food. Its energy is obtained by oxidative phosphorylation and the muscle mass has abundant mitochondria and plenty of myoglobin—these two giving it its colour. (What colour muscle would you expect to find in the breast of a bird capable of prolonged flight?)

Flying insects tend not to operate anaerobically. They would not be able to carry the large quantities of fuel required to sustain a long flight powered by anaerobic glycolysis.

Diving mammals have muscle which is strongly red in colour because of the large amounts of myoglobin present. *Diving reptiles* produce large quantities of lactate, but are capable of withstanding the comparatively large fall in the blood pH (pH 7.4 to pH 6.7) which results. This taken together with other factors such as a decrease in the heart rate, allows freshwater turtles to survive 12 hours without oxygen. The *brains* of all animals are entirely dependent upon oxidative phosphorylation for their energy supply—and hence upon oxygen. In diving mammals a special mechanism (reflex) allows the blood supply to skeletal muscles to be shut off while that to the brain is maintained.

High altitudes. Many millions of people live and work at high altitudes (4 000 m above sea-level) where the partial pressure of oxygen in the atmosphere is nearly half that at sea-level. Individuals who work at such altitudes show adaptations including an increase in the number of red blood corpuscles in the blood (blood haemoglobin concentration 23 g dl^{-1} compared with a normal 15 g dl^{-1}).

Modern airliners fly at about 10 000 m, and Concorde at 16 000 m. Passengers would normally lose consciousness and die in a few minutes at these altitudes where the oxygen concentration may be only one-tenth of that at sea-level. Therefore aircraft cabins are pressurized, usually to a pressure equivalent to that at about 2 600 m. To pressurize to that at sea-level would demand much greater cabin strength and a prohibitive weight increase.

Do all anaerobic organisms form lactate?

Some bacteria run their affairs in a similar way to vertebrate muscle, producing lactate. However, other organisms faced with the same problem of regenerating NAD$^+$ from NADH, make ethanol instead. Yeast is the best-known example here and 'alcoholic fermentation' is an enormous industry in many countries of the world.

In yeast, in anaerobic conditions, the pyruvate formed as a result of glycolysis, is first converted to acetaldehyde (ethanal) by the loss of carbon dioxide (Figure 3.35). The acetaldehyde then reacts with NADH in an enzyme-catalyzed reaction, to give ethanol and NAD$^+$. The yeast can do nothing with the ethanol it has produced and gets rid of it into the surrounding medium. Humans drink the resulting dilute solution of ethanol.

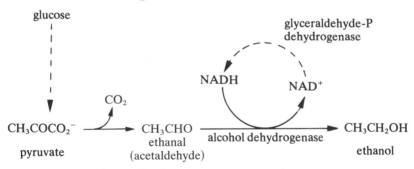

Figure 3.35 Alcoholic fermentation in yeast

Mini-cycles

We may pause here to reflect that coenzymes such as NAD$^+$ are only present in small amounts and are constantly being recycled: NAD$^+$⇌NADH. There would be no point in accumulating NADH. A similar situation applies to other coenzymes and to ATP. ATP and ADP are constantly being recycled: ADP⇌ATP, and the total amount of ADP *plus* ATP in a cell is quite small. ATP is made *when* it is needed *where* it is needed.

Converting pyruvate to acetyl coenzyme A

In aerobic conditions these problems with coenzymes do not arise, and pyruvate from glycolysis is converted to acetyl coenzyme A in both vertebrate muscle and yeast cells. The reaction basically involves the C_3 compound, pyruvate, losing carbon dioxide to form a C_2 acetyl unit:

$$CH_3COCO_2^- \xrightarrow{\quad CO_2 \quad} CH_3CO-$$

pyruvate   acetyl unit

During this transformation some energy is released, and living organisms do not pass up the opportunity of trapping this energy as ATP.

The conversion of pyruvate to acetyl coenzyme A, although in principle a simple reaction, is catalyzed by an immensely complicated enzyme. The 'enzyme' is really a complex of several enzyme proteins which form an aggregate, and the complex requires two vitamins of the B group (Box 3.9) for activity, NAD^+ (which is converted to NADH) and, or course, coenzyme A. The NADH formed can give rise to ATP by oxidative phosphorylation: the acetyl coenzyme A formed can feed into the TCA cycle, again with the formation of ATP.

In this way, by 'aerobic glycolysis', glucose is *completely* catabolized to carbon dioxide and water. In the first phase, as far as pyruvate or lactate, little ATP is generated, but that which is, is generated anaerobically. In the second phase, pyruvate is broken down to carbon dioxide and water with the aerobic release of large amounts of energy as ATP.

It is possible to draw up a balance sheet of the ATP used and made during the overall sequence of glycolysis and TCA cycle.

	Potential ATP: Debit	Credit
(Starting material: glucose. Debits and credits expressed per glucose molecule)		
Activation to give C_6 diphosphate	2	–
Substrate level phosphorylations ($\times 2$)	–	4
2 NADH formed in glyceraldehyde 3-phosphate step (Figure 3.34)	–	6
Pyruvate → Acetyl CoA ($\times 2$)	–	6
Acetyl CoA → TCA Cycle ($\times 2$)	–	24
Net yield of ATP		38

Box 3.9 The vitamins

Humans, and other animals, require certain organic compounds to be present in the diet in small but adequate quantities. If these compounds are not present then deficiency symptoms and death may result.

The substances required in this way are referred to as 'vitamins', and the term covers a wide range of chemically unrelated compounds which perform a wide range of functions in metabolism. They are distinguished from the bulk nutrients (carbohydrates, fats, proteins) which supply energy or raw materials, in being needed only in very small amounts. The average daily requirement for humans is in the microgram to milligram range, depending on the vitamin. The other feature that distinguishes them is that animals cannot manufacture them and must obtain them in the preformed state. They are, of course, synthesized by plants, and in many instances by micro-organisms. Indeed bacteria in the gut often supply a good deal of the vitamins required by an animal, reducing or abolishing the dietary requirements.

For the majority of the vitamins, their function in metabolism is known. The compounds were previously designated by letters (vitamin A, B, C, and so on) before their chemical structures were known. Now that the chemical structures *are* known, the letter designations are often still used, being simpler than the long chemical names.

Fat-soluble vitamins. The vitamins A, D, E and K are insoluble in aqueous solvents and are referred to as 'fat-soluble vitamins'. Each has its own distinct function in the body, and the only reason for grouping them together is their solubility properties. Their overall functions are known, but somewhat less is understood of their detailed mode of action than of the second group of vitamins, the 'water-soluble vitamins', discussed below. The table below gives some information about the chemical nature, function and dietary sources of the fat-soluble vitamins.

Vitamin	Chemical nature	Dietary source	Function
A	Retinol (C_{20} unsaturated alcohol)	Dairy produce, eggs, liver. Animals can make it from plant carotenes	In vision as prosthetic group to the protein rhodopsin. Other functions, too, e.g. maintenance of epithelial cells
D	Cholecalciferol = sterol	Dairy produce, yeast. Can be made in human skin by action of sunlight	In promoting absorption of calcium from intestine and hence normal formation of bone
E	α-Tocopherol	Vegetable oils, liver, eggs, milk, butter	Several depending on species, e.g. prevention of sterility in rats. Functions in humans uncertain
K	Naphthoquinone derivatives	Green leaves, intestinal flora	Required for normal blood clotting

Vitamin C is a water-soluble carbohydrate called ascorbic acid, abundant in fresh green vegetables and citrus fruit. Most animals can make their own ascorbic acid. However, man, monkeys and guinea-pigs cannot, and the human daily requirement is 30–50 mg. Deficiency of vitamin C causes *scurvy*, characterized by fragility of the small blood vessels leading to bleeding into the tissues. Exactly how this arises is not certain, but it is known that ascorbic acid plays a critical role in the formation of connective tissue, especially collagen.

The B-group vitamins. Practically all the other water-soluble vitamins fall into the 'B group', having been at one time or another been given a 'B' classification, e.g. 'vitamin B_6'. Their chemical structures are well known and almost all play a role as coenzymes in various parts of metabolism. Many of the coenzymes with which we are familiar, and which function in a multitude of reactions (e.g. NAD^+), are formed from B-group vitamins.

These water-soluble vitamins are, of course, present in practically all living materials, but particularly rich sources are cereals and yeast.

Thiamine (previously vitamin B_1) is required for the intracellular synthesis of thiamine pyrophosphate, a coenzyme participating in the reactions by which pyruvate is converted to acetyl coenzyme A (p. 134) and α-ketoglutarate is converted to succinate in the TCA cycle (Fig. 3.28). Deficiency of thiamine results in a disease called *beriberi.*

Niacin, or nicotinic acid, forms part of the molecules NAD^+ and $NADP^+$, which participate in a large number of dehydrogenase reactions. It is not surprising that deficiency results in a number of symptoms, the disease *pellagra* in humans, and *black tongue* in dogs. Humans can produce nicotinic acid from the amino acid tryptophan, but the normal requirement is not fulfilled by the average tryptophan intake (in excess of what is required for protein manufacture).

Riboflavin (previously vitamin B_2) forms part of the flavin coenzymes (FMN, FAD) which participate in a large number of dehydrogenation reactions, especially those going on in the mitochondria.

Pyridoxine (previously vitamin B_6) has a number of diverse roles as a coenzyme. A well-known function is as a coenzyme of transaminases. The drug *isoniazid*, used in the treatment of tuberculosis, produces pyridoxine deficiency by combining with the vitamin.

Biotin is involved in a number of enzyme-catalyzed reactions in which carbon dioxide is added to a compound, e.g. the formation of malonyl coenzyme A in the biosynthesis of fatty acids.

Pantothenic acid is required as part of the structure of coenzyme A itself.

Folic acid, and tetrahydrofolic acid, play vital roles in the interconversion of certain amino acids and in the synthesis of nucleotides, especially when C_1 units are carried from molecule to molecule.

Cobalamin (previously vitamin B_{12}) is a complicated porphyrin-like compound containing the metal cobalt. It does not occur in plant foods and is synthesized exclusively by micro-organisms. The recommended daily requirement for a human is 3 µg. The vitamin has several functions, one of which is in conjunction with folic acid in transferring C_1 units. *Pernicious anaemia* results, in humans, when the gastric cells fail to secrete a glycoprotein called 'intrinsic factor' essential for absorption of cobalamin from the diet. The symptoms may be relieved by intravenous *injection* of the vitamin.

Other B-group vitamins are lipoic acid, choline, myo-inositol, and ubiquinone, but do not appear to be essential under all conditions. Some can be synthesized provided certain precursor molecules are present in the diet.

3.8. Protein breakdown

In addition to carbohydrate and fat, the diet also contains proteins, and western diets are especially rich in protein. Dietary protein has two functions. One is to supply the animal body with certain of the amino acids which it is incapable of making. About 9 or 10 of the amino acids in proteins are classified as *essential* to humans: if they are not present in the diet, deficiency symptoms and eventually death occur (p. 241). Only comparatively small amounts of protein containing all the essential amino acids are needed for this purpose, although a young and growing animal will have greater need in this respect than an adult.

Any protein in the diet in excess of that needed by an animal to build up its own types of protein is catabolized: the amino acid residues are oxidized to obtain energy.

Is there a store of protein comparable to glycogen and fat in animals, or starch in plants? Some plant seeds have protein stores, and milk protein (casein) represents a way of passing amino acids from mother to infant, but in the main animals do not have proteins whose only function is to act as stores. However, when an animal is starving, and when the glycogen and fat stores are exhausted, then body protein starts to be broken down to be used as food, that is, as an energy source. This is a last resort, and is like burning the furniture in order to keep warm. Nevertheless it does happen; mainly the muscle proteins are used and we see 'wasting'.

How are amino acids catabolized?

Dietary proteins are broken down to amino acids by the pepsin in the stomach, and by trypsin and chymotrypsin and other enzymes in the small intestine. These enzymes have a range of specificities as to which peptide bonds between which particular amino acid residues they will split: the result is that practically all proteins are broken down to amino acids regardless of the original amino acid sequence. These amino acids pass through the intestinal epithelium into the blood and hence reach the liver. From here they can be distributed, in the blood, to organs in need of amino acids for building proteins: otherwise they are catabolized, mainly in the liver.

The breakdown of amino acids occurs by a number of routes—as befits a heterogeneous group of molecular structures. We will concentrate on a couple of common features only.

The first common factor for all amino acids is the need to get rid of the nitrogenous part of the molecule ($-NH_2$). In mammals, the nitrogen is

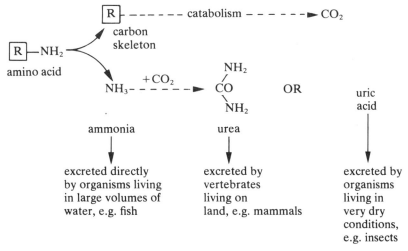

Figure 3.36 Disposal of excess nitrogen from dietary amino acids and proteins used as energy sources rather than raw materials. Ammonia is highly toxic to all organisms. Animals living in large masses of water excrete ammonia directly trusting that it will be diluted out by the water in which they live. Both urea and uric acid are much less toxic. It does however require the expenditure of energy to make them. Uric acid is very insoluble in water: crystals of uric acid can be excreted with practically no water loss

excreted in the urine as urea. A sequence of metabolic reactions exists for this conversion, called the Urea Cycle. *This* cycle was discovered by Krebs about 1934—*before* the TCA cycle had been found.

Other animals, such as fish, excrete their excess nitrogen in the form of *ammonia*. Because ammonia is highly toxic to all animals, it is necessary to get rid of it promptly, and in large volumes of water, and so excreting ammonia is only satisfactory for animals living in water. In contrast, animals living in conditions of acute water shortage make another nitrogenous compound, *uric acid*, which is less toxic than urea and very much less toxic than ammonia, and is practically insoluble. Insects and birds fall into this category. These routes of nitrogen excretion are summarized in Figure 3.36.

The part of the amino acid remaining after the amino group has been removed (i.e. after deamination) is called the 'carbon skeleton', but, of course, this skeleton will have all sorts of shapes depending on what the particular amino acid was. There are therefore a number of different catabolic pathways for dealing with these skeletons. Commonly the skeleton, or parts of it, are converted to acetyl coenzyme A molecules, or alternatively to molecules that can be fed into the TCA cycle (e.g. pyruvate, acetate, oxaloacetate). The amino acid alanine, for example, yields pyruvate (Figure 3.37), and hence acetyl coenzyme A.

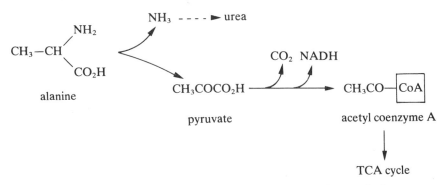

Figure 3.37 Metabolic route for the disposal of excess dietary alanine

Summary of catabolic processes

It is now possible to take a step backwards to survey the multitude of metabolic pathways we have discussed in the foregoing account. Taking a less detailed view, we can discern the overall picture of the 'metabolic funnel' for dealing with thousands of different types of foodstuffs in a very economical way. We can also see the 'aim' of catabolism, that is, to 'juggle' with the atoms making up organic compounds to get them into a form suitable for using in the ATP-making processes of the cell. These relationships are summarized in Figure 3.38. Only one thing is missing. In such a set of complex interrelationships there must be some sort of *control*. For example, how is it decided, during a period of fasting, whether glycogen or fat should be utilized as an energy source, and how much should be broken down to meet the energy needs of the organism?

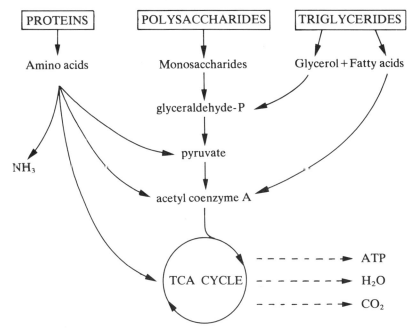

Figure 3.38 Summary of catabolic pathways

3.9. Control of metabolism

In addition to the pathways of metabolism outlined in this chapter there are many many more pathways such as those responsible for *biosynthesis* of all the cell's component compounds. It is extremely important that the right things are broken down at the right times, and the correct chemicals synthesized at the correct locations, and so on. For instance, it would be most unfortunate if the metabolic system of an animal catabolized glycogen to obtain energy in the form of ATP and used the ATP simultaneously to lay down a store of glycogen. This would simply waste ATP! The need for a tight control on what happens in metabolism is obvious.

How can sequences of reactions be controlled?

The metabolic pathways in a cell are multi-step sequences of reactions catalyzed by enzymes. When the pathway is operating, the reactions are all in a steady-state condition, not at equilibrium. Therefore it is in principle possible to control the pathway simply by turning off one step in the sequence: this is something like having a tap in a water-line. Can the catalytic activity of the enzyme molecules be 'turned down' or 'turned up'? Not really: either catalysis takes place or it doesn't. However, catalytic activity can be 'turned off' or 'turned on' in a number of ways, achieving the same ends. All of these ways involve making a small and reversible chemical or physical change to the enzyme molecule. We shall now look at two examples where the control is achieved in different ways. As it happens both have to do with carbohydrate metabolism, but in fact all the other pathways are just as tightly controlled. A more radical way of changing enzyme activity is for the cell to make more enzyme-protein molecules. This does happen, but it is more convenient to discuss it at the end of Chapter 5 when we have discussed the biosynthesis of proteins.

Self-control of glucose metabolism

It was observed a long time ago that a suspension of yeast cells consumed much more glucose if it was respiring anaerobically rather than aerobically. We can easily explain this from what we already know. Under anaerobic conditions the yeast cells break down glucose via the glycolysis pathway to produce, finally, ethanol. Large amounts of glucose have to be passed through the pathway in order to get a comparatively small amount of ATP. On the other hand, under aerobic conditions ethanol is not produced, but instead the end product of glycolysis, pyruvate, is converted to acetyl coenzyme A which enters the TCA cycle. Large amounts of ATP are produced for each molecule of glucose metabolized in this way (Figure 3.39). It is clear that to get the same amount of ATP, much more glucose must be metabolized anaerobically

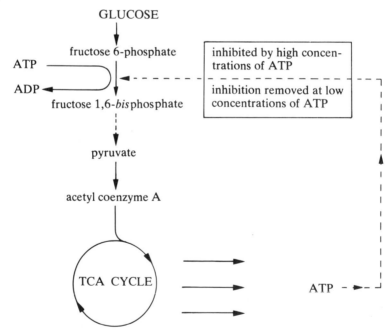

Figure 3.39 Feedback control of the amount of glucose passing through glycolysis and the TCA cycle. The TCA cycle produces very much more ATP per glucose molecule than does the glycolysis pathway. The operation of the kinase enzyme that converts fructose phosphate to fructose *bis*phosphate (fructose diphosphate) is controlled allosterically by the prevailing ATP concentration. When the TCA cycle is operating, i.e. under aerobic conditions, the ATP concentration tends to be relatively high. This enzyme therefore tends to be inhibited, and therefore relatively less glucose passes through the pathway as a whole. The opposite happens in anaerobic conditions

than aerobically. What actually happens is that the yeast cells turn down the rate of operation of the glycolysis pathway in the presence of oxygen so that less glucose is used. How does this come about?

It is possible to identify a single enzyme-catalyzed step in glycolysis as the controlling one. This is the step where a second phosphate is added to the sugar molecule to give a diphosphate (actually fructose with two phosphates attached). The enzyme that catalyzes this step, a *kinase*, is inhibited by high concentrations of ATP as well as by certain other products of metabolism. When the TCA cycle and oxidative phosphorylation are taking place, i.e. in aerobic conditions, a great deal of ATP is produced and its concentration in the cell rises. Hence the kinase enzyme is inhibited and the result is that the amount of glucose flowing through the glycolysis pathway is diminished. Under anaerobic conditions the ATP concentration falls because no oxidative phosphorylation is going on, and the rate of glycolysis is restored. The system is therefore self-regulating and responds to the prevailing conditions, whether aerobic or anaerobic.

The phenomenon observed here is that an end product of a pathway—in this case ATP—inhibits a step quite early in the sequence of reactions (Figure 3.39). This is called *feedback inhibition*. It is obviously sensible to inhibit the sequence early on to avoid making a lot of intermediates in the pathway which are of no use for anything else.

The inhibition of the kinase enzyme mentioned above by ATP (and other metabolites) happens because the conformation of the enzyme-protein is changed by the presence of relatively-high concentrations of ATP. This is *allosteric inhibition*, mentioned earlier (p. 73). Many examples of this type of inhibition and feedback control are known. Furthermore, although the example given above was yeast cells, glucose utilization is 'self-controlled' in this way by almost all organisms, not just yeasts. It means that the right amount of glucose is used for the needs of the cell. Any situation in which ATP is used results in a lowering of the ATP concentration, which in turn means that the kinase enzyme becomes less inhibited and more glucose flows through the pathway.

Control of glycogen utilization

It is obviously very important to an animal to control the size of its glycogen 'stock'. When no glucose is available from the diet, and when energy is needed for the contraction of muscles, glycogen needs to be broken down to provide glucose. After a meal, on the other hand, is an appropriate time for building up glycogen stores, not breaking them down. How is this control achieved?

In this case the system is not entirely self-controlling, but is regulated from without the cell by *hormones*. A hormone is a chemical substance produced in a ductless gland in the body which circulates in the blood and 'influences' a *target organ*. The chemical nature of hormones is extremely varied, but as a class of 'biological agents' they are characterized by being active in minute quantities. Several hormones control carbohydrate metabolism: in this section we will discuss the action of the hormone *adrenaline* on glycogen breakdown.

Adrenaline (also called epinephrine) is produced by the adrenal glands situated near the kidneys. It takes part in the so-called 'fight or flight' reaction of an animal to imminent danger. The brain perceives a dangerous situation about to arise and sends a message along nerves to the adrenal gland. The glands are then stimulated to secrete adrenaline into the bloodstream. The consequences of this are manifold, the animal being 'tensed up' ready to fight or run for its life—adrenaline affects many parts of the body. It is clear, however, that whatever happens, vigorous muscular exercise is likely to be called for, and hence an energy supply will be required. It is appropriate therefore to convert glycogen quickly to glucose ready for glycolysis. How does adrenaline achieve this?

The detailed story is quite complicated but as with most biochemical mechanisms the principles are simple and straightforward. Glycogen breakdown involves removing glucose residues one by one from the branched-chain polymer. The removal is essentially a reversal of the condensation reaction in which glycogen was formed, and so *should* involve the addition of a molecule of water. However, instead of water (H—OH), phosphoric acid (H—PO$_4$H$_2$) is used, and the product is glucose phosphate which can feed directly into the glycolysis pathway. (This incidentally saves using one molecule of ATP compared with when glucose itself is used.) The enzyme responsible for catalyzing

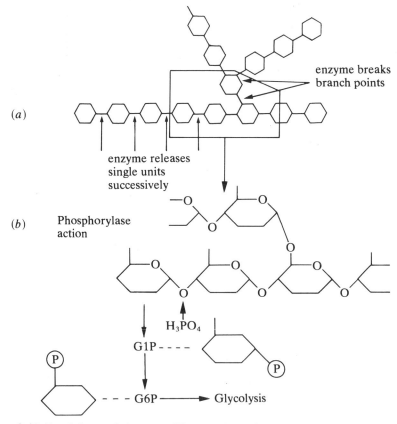

Figure 3.40 Breakdown of glycogen. Glycogen has $\alpha 1 \rightarrow 4$ links between the glucose residues, with $\alpha 1 \rightarrow 6$ links forming branch points. The enzyme glycogen phosphorylase catalyzes the cleavage of the $\alpha 1 \rightarrow 4$ links: other enzymes, including 'debranching enzyme' are required to deal with the branch points

the addition of phosphoric acid is called *glycogen phosphorylase*. Other enzymes are needed additionally to cope with branch points in the glycogen molecule, but these need not concern us here (see Figure 3.40). It is the phosphorylase enzyme that is controlled, and the adrenaline that does the controlling. Strangely however adrenaline never enters the cell, but is bound to a special receptor in the cell membrane. When this happens changes are triggered off in the membrane resulting in the formation of a compound called *cyclic AMP* from ATP, on the *inside* of the membrane. Cyclic AMP is sometimes referred to as the 'second messenger', the first messenger being the hormone, adrenaline in this case.

Once formed cyclic AMP causes a sequence of changes in the cell which results in the turning on of the catalytic activity of an enzyme whose function is to add a phosphate group to the protein of glycogen phosphorylase. This phosphate group is supplied by ATP. The key point is that glycogen phosphorylase is *only* active in breaking down glycogen when it is phosphorylated in this way. Hence adrenaline activates glycogen breakdown.

This is a complicated sequence (Figure 3.41) to do an apparently simple job, and there is yet another sequence of reactions to *remove* the phosphate group from the protein so that the effect of the adrenaline 'wears off' after a short time. Probably one reason for having a complicated sequence like this is that it *amplifies* the response. One molecule of adrenaline leads to the production of several dozen molecules of cyclic AMP, and each of these leads to the activation of several thousand molecules of glycogen phosphorylase.

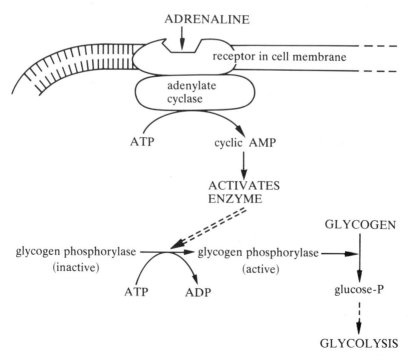

Figure 3.41 Adrenaline activates glycogen breakdown by a cascade system

Thus an amplification of some ten thousand times occurs because at each step *a catalytic activity* is turned on. This is why adrenaline, and indeed other hormones too, are effective in such small amounts. Also, the response is rapid: the muscle obtains an energy supply quickly.

Box 3.10 Diabetes

The disease diabetes mellitus, usually just 'diabetes', is characterized by raised levels of blood glucose as well as increased fat and protein metabolism. It is one of the most prevalent of human metabolic defects. About one person in every 2 500 will develop the so-called juvenile diabetes between the ages 8 and 12, and about 3 in every 100 develop the maturity onset type of the disease between the ages 40 and 50. Over 7% of people in their late 70s are affected. There are genetic factors involved, but these do not behave in a straightforward way.

The disease varies greatly in severity. In many instances it can be treated successfully by regulating the diet alone; otherwise it is necessary to give daily injections of insulin or treat with sulphonylurea drugs such as 1-butyl-*p*-tosylsulphonylurea (Tolbutamide):

$$CH_3 \text{—}\!\!\left\langle \bigcirc \right\rangle\!\!\text{—} SO_2-NH-CO-NH-C_4H_9$$

There are in fact several 'causes' of the disease. Failure of the β-cells of the Islets of Langerhans of the pancreas is one cause, but in many adult diabetics the problem is a decreased sensitivity to insulin. This latter class of patients responds to the sulphonylureas. In other patients the problem may be a too rapid destruction of insulin.

The most easily recognizable symptom in diabetics is a very high blood glucose level. The level in normal individuals is about 5 mM, but in untreated diabetics it may be 8–60 mM. This results in glucose appearing in the urine, and a characteristic of the disease is the production of large volumes of urine. (The word 'diabetes' is Greek for 'siphon'!)

The lack of proper feedback control of blood glucose level and the failure of blood glucose to enter cells, results in excessive synthesis of glucose by certain tissues including the liver. Since carbohydrate cannot be formed from fat, metabolism is switched to breaking down proteins, either dietary or body, to provide amino acids which can be converted to glucose. Liver glycogen is depleted, excess nitrogen from protein degradation appears in the urine, and fat degradation continues with the accumulation of the degradation products known as ketones.

Despite a great deal of research work over the years since insulin was isolated in 1921 by Banting and Best, it is still uncertain how insulin achieves its effect. It is agreed, however, that insulin action results in permeability changes in cell membranes: presumably membranes have receptors to which the insulin molecule attaches. It has been suggested that a compound, cyclic GMP, analogous to cyclic AMP, acts as the 'second messenger' in insulin action.

Insulin for the treatment of diabetic individuals is obviously required in considerable amounts. It is normally isolated from the pancreas of cattle or pigs. It is possible to synthesize insulin chemically but yields are very poor because of the difficulties of producing disulphide bonds in the correct places. In recent years, there has been great interest in synthesizing insulin using the techniques of 'genetic engineering'. This involves introducing the gene for insulin into a micro-organism, growing the organism in culture and isolating the insulin produced (see Fig. 6.4).

Other controls on carbohydrate metabolism

Glucose utilization in the vertebrate body is controlled by two polypeptide hormones from the endocrine pancreas, *insulin* and *glucagon*. Despite a great deal of intense research activity, we do not have anything like as clear a picture of how they act as we do with adrenaline.

Insulin is produced by the β-cells of the *Islets of Langerhans* of the pancreas in response to an elevation in the concentration of glucose in the blood. Deficiency of insulin causes the disease *diabetes* (Box 3.10). Insulin is known to have a number of actions in the body. For example, it stimulates, possibly by altering membrane permeability, uptake of glucose from the blood by muscle and adipose cells. It has no such effect on liver cells, but this makes sense if we look at the relationships between the different tissues. If the blood going to the liver (e.g. in the portal vein) is rich in glucose, the liver removes some of it for conversion and storage as glycogen. When the blood glucose level falls the liver breaks down glycogen to glucose and secretes this into the blood. Other tissues such as muscle and adipose tissue then pick up this glucose.

Insulin also appears to promote the catabolism of glucose *and* the incorporation of glucose into liver and muscle glycogen. These actions too tend to reduce the blood glucose level. It also has effects on fat and protein metabolism and these effects are distinct from those on cell membranes.

Glucagon is produced by the α-cells of the Islets of Langerhans. It has the opposite effect of insulin, that is, it causes an increase in the blood glucose concentration.

It will be seen that carbohydrate metabolism is very tightly controlled by a number of mechanisms which interlock to form a complicated system to maintain homeostasis. It is difficult to study the effects of, say, one hormone in isolation. Nevertheless for those hormones we do understand at the present time, it is perfectly possible to give a *chemical explanation* of their mode of action. This is, of course, the biochemist's aim, to provide chemical explanations for biological phenomena.

Questions

1. With the aid of labelled drawings, describe the structure of (*a*) a mitochondrion, and (*b*) a chloroplast.
   Compare the functions of these two organelles.   [O & C]

2. (*a*) Explain, in outline only, how ATP (adenosine triphosphate) is required for and generated by aerobic respiration in a plant cell.
   (*b*) Outline the role of ATP in the process of photosynthesis.
                                                                       [JMB]

3. (*a*) Why is ATP so important a compound?
   (*b*) Give an account of *two* processes in which ATP is essential.
   (*c*) Outline *two* different ways by which cells can produce ATP.
                                                                       [O & C]

4. Review the fate of one molecule of glucose during complete respiratory catabolism.

5. Mammalian liver cells were broken up (homogenized) and the resulting homogenate centrifuged. Portions containing only nuclei, ribosomes, mitochondria and residual cytoplasm were each isolated. Samples of each portion, and of the complete homogenate, were incubated in four ways: with glucose, with pyruvic acid, with glucose and cyanide, and with pyruvic acid and cyanide. (NB. Cyanide affects the oxidation of cytochromes.) After incubation the presence or absence of CO_2 and lactic acid in each sample was ascertained. The results are summarized in the table.
   (a) (i) Where in the cell would you expect to find the enzymes of the Krebs cycle? Explain how the results in the table support your answer.
      (ii) Explain why cyanide stops the production of CO_2.
   (b) (i) Which fraction of the homogenate contains the enzymes able to convert pyruvic acid to lactic acid?
      (ii) Explain *in one sentence* why cyanide has no effect on the lactic acid production from both glucose and pyruvic acid.
   (c) Explain why CO_2 can be produced from glucose by the homogenate, when none of the separated portions do this.

Incubated with	Complete homogenate		Nuclei only		Ribosomes only		Mito-chondria only		Residual cytoplasm only	
	CO₂	LA	CO₂	LA	CO₂	LA	CO₂	LA	CO₂	LA
Glucose	√	√	×	×	×	×	×	×	×	√
Pyruvic acid	√	√	×	×	×	×	√	×	×	√
Glucose + cyanide	×	√	×	×	×	×	×	×	×	√
Pyruvic acid + cyanide	×	√	×	×	×	×	×	×	×	√

LA = lactic acid   √ = presence   × = absence

   (d) Which *three* of the following would you expect to be particularly rich in mitochondria:
      (i) muscle cell,
      (ii) adipose tissue cell,
      (iii) xylem vessel,
      (iv) liver cell,
      (v) red blood cell,
      (vi) apical meristematic cell?   [JMB]

6. The scheme below shows the relationships of various metabolic pathways:

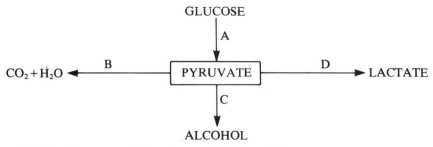

   (a) Give the names of the pathways A, B, C and D.

(b) Indicate the approximate amount of ATP generated by pathways A + B with a tick.

(i) 4 mol/mol glucose   (iii) 400 mol/mol glucose
(ii) 40 mol/mol glucose   (iv) 4 000 mol/mol glucose

(c) In which of the pathways A–D does acetyl coenzyme A occur?

(d) Which of the end products is equivalent to half a glucose molecule?

7. This is a diagram of the internal structure of a chloroplast as drawn from an electron micrograph.

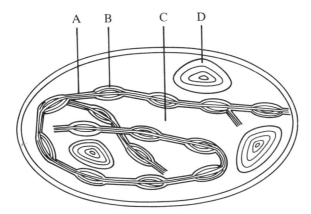

(a) Name the structures A, B, C, D.

(b) At which one of A–D does the light dependent stage of photosynthesis occur?

8. In an experiment with mitochondria, a medium which contained inorganic phosphate and oxidizable substrates was used. The medium was saturated with air at the start. The experiment determined the fall in oxygen concentration in the medium when firstly mitochondria, and then ADP, were added.

The letters used below refer to those on the graph.

In the experiment, mitochondria were added at W. A standard amount of ADP was added at X, a similar amount at Y, and at Z.

In a control experiment, mitochondria were added at W. No ADP was added.

The results are shown in the graph.

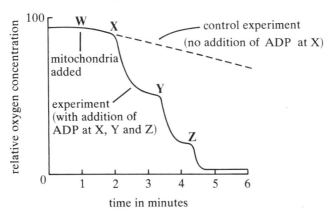

(a) State *one* function of the phosphate in the medium.

(*b*) Account for:
  (i) The slight drop in oxygen concentration when the mitochondria were added at W,
  (ii) The shape of the graph when the ADP was added at X and Y.
(*c*) Why does the graph become horizontal shortly after the addition of ADP at Z?
(*d*) Explain briefly what effect the following additions might have:
  (i) The addition of excess ADP at X.
  (ii) The addition of cyanide at X as well as excess ADP.

[JMB]

9. (*a*) The following equation may be used to summarize the chemical and energy exchanges which occur in respiration. Complete the equation by inserting the names (or abbreviations) and number of molecules of the missing compounds.

$$C_6H_{12}O_6 + 6O_2 + \cdots + \text{phosphate} \rightarrow 6CO_2 + 6H_2O + \cdots$$

(*b*) The first stage in respiration involves the conversion of one molecule of glucose into two molecules of a three-carbon compound. State: (i) the name of this stage; (ii) where this stage occurs in the cell; (iii) the name of the three-carbon compound formed; (iv) how much energy is conserved as ATP during this stage.
(*c*) If oxygen is present other events occur within the cell. The three-carbon compound enters a specific cell organelle and a six-carbon compound is formed. This compound is in turn gradually broken down. State: (i) the name of the organelle; (ii) the name of the six-carbon compound.
(*d*) During the breakdown, protons and electrons are gradually released. What important event accompanies this release of protons and electrons? What substance acts as the final hydrogen acceptor in the organelle? Name one substance involved in the transfer of hydrogen to the final acceptor.

[JMB]

Further reading

Chappell, J. B. (1977), 'ATP', *Oxford/Carolina Biology Reader*, **50**.
Chappell, J. B. (1978), 'Energetics of Mitochondria', *Oxford/Carolina Biology Reader*, **19**, (Revised Edition).
Govindjee, and Govindjee, R. (December 1974), 'The Primary Events of Photosynthesis', *Scientific American*, **231**, 68.
Hall, D. O., and Rao, K. K. (1977), 'Photosynthesis', *Studies in Biology*, **37**, (Second Edition). (Edward Arnold).
Hinckle, P. C., and McCarty, R. E. (March 1978), 'How Cells Make ATP'. *Scientific American*, **238**, 104.
Jones, C. W. (1976), 'Biological Energy Conservation'. (Chapman & Hall).
Miller, K. R. (October 1979), 'The Photosynthetic Membrane', *Scientific American*, **241**, 102.
Rabinowitch, E. I., and Govindjee (July 1965), 'The Role of Chlorophyll in Photosynthesis'. *Scientific American*, **231**, 74.
Stoeckenius, W. (June 1976), 'The Purple Membrane of Salt-Loving Bacteria', *Scientific American*, **234**, 38.
Tribe, M., and Whittaker, P. A. (1972), 'Chloroplasts and Mitochondria', *Studies in Biology*, **31**. (Edward Arnold).
Walker, D. (1979), 'Energy, Plants and Man'. (Packard Publishing).
Whatley, J. M., and Whatley, F. R. (1980), 'Light and Plant Life', *Studies in Biology*, **124**. (Edward Arnold).
Whittingham, C. P. (1971), 'Photosynthesis', *Oxford/Carolina Biology Reader*, **9**, (Second Edition).

4
Using
Energy

Summary

The energy (ATP and reducing power) obtained either by trapping sunlight or by oxidizing food materials is used by living organisms to drive all their various life activities. Photosynthetic organisms make all of their cellular constituents from simple compounds such as carbon dioxide, water and ammonia. They are dependent upon micro-organisms, both in the soil and symbiotic, for the fixation of atmospheric nitrogen as ammonia. Reduction of carbon dioxide to organic compounds [$(CH_2O)_n$], and of nitrogen to ammonia, requires reducing power: the formation of new chemical bonds is driven by coupling biosynthetic reactions to the hydrolysis of ATP. All organisms store surplus energy and raw materials for times of shortage, usually as either carbohydrate (starch, glycogen) or fat or both. Biosynthetic sequences of reactions usually differ from breakdown sequences in detail, so that they may be controlled separately. Biosynthesis of cellular constituents is but one of the ways in which energy is used by living organisms. There are many others, the most studied being muscular contraction whereby the chemical energy released by ATP hydrolysis is transformed into mechanical energy.

4.1 Introduction

The major theme of Chapter 3 was how organisms obtain energy either by means of the light reaction of photosynthesis or by the catabolism of food and storage materials. Both of these processes result in the production of reduced coenzymes such as NADH or NADPH, and the formation of ATP from ADP and inorganic phosphate. Chapter 4 is about how these two 'forms' of biologically-usable energy drive energy-requiring processes such as chemical synthesis and movement.

Chemical synthesis in biological organisms, or 'biosynthesis', is necessary for growth, the ability of living things to build more of their own substance, and reproduction, forming new individuals of a species. Biosynthesis, basically joining atoms together, is a process that requires chemical energy. As well as growth, maintenance is necessary—replacement or repair of damaged parts—but this is essentially the same as growth except that no net change in cell mass occurs.

149

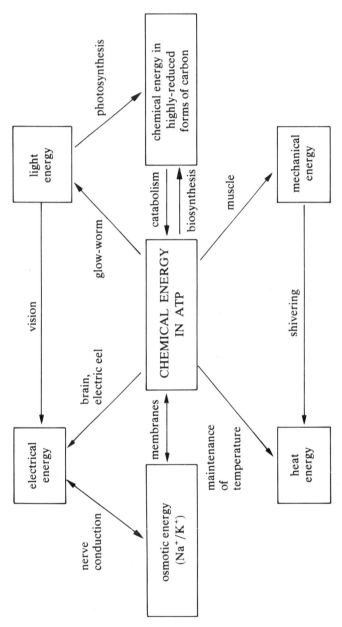

Figure 4.1 Energy transformations in living organisms

Muscular contraction involves converting stored chemical energy to mechanical energy for movement. There are many other 'energy transformations' that occur in organisms such as ciliary and flagellar movements, producing light in glow-worms, producing an electric current in electric eels and performing osmotic work, e.g. moving materials *against* concentration gradients. All of these may be powered by processes involving the hydrolysis of ATP (Figure 4.1).

Building blocks and information

Growth and maintenance, that is, building more organic substance, requires: (1) *the building blocks* with which to make macromolecules including monosaccharide sugars, amino acids and nucleotide bases to make polysaccharides, proteins and nucleic acids, respectively; and (2) *information*, so that the macromolecules characteristic of a given organism are produced. Plants build cellulose from glucose to incorporate into their own type of cell walls. Insects build chitin from other monosaccharide sugars to incorporate into their own characteristic types of exoskeleton. These characteristic materials are formed because plants, for example, have cellulose-synthesizing enzymes and not chitin-synthesizing ones. Eventually the genes 'decide' which types of enzymes are made in a given cell—but this is a matter for Chapter 5. We merely summarize here by saying that there is an input of *information* leading to the synthesis of the 'right' types of enzyme in a particular cell type.

We now return to ask where the building blocks come from. Photosynthetic organisms produce their own building blocks: they make sugars, amino acids, nucleotide bases, as well as fatty acids, from carbon dioxide, water and ammonia. Heterotrophs obtain their building blocks by consuming plants or other heterotrophs, i.e. their building blocks are preformed. All organisms, photosynthetic and heterotrophic, join up building blocks to make their own characteristic macromolecules.

We are going to consider first how plants build up their building blocks, using NADPH and ATP formed in the light reaction of photosynthesis, in the so-called *dark reaction* of photosynthesis. It is important to realize that *all* the organic compounds a plant is made up from are synthesized by the plant—not just carbohydrates, although carbohydrates may be the initial products.

4.2 The dark reaction of photosynthesis

Earlier (p. 101) we described the process of converting carbon dioxide to the organic constituents of plants as requiring two things: the reduction of carbon dioxide (CO_2) to (CH_2O), and the linking of carbon atoms together to form carbohydrates, amino acids and fatty acids, etc.:

$$CO_2 \xrightarrow{\text{reduction}} (CH_2O) \xrightarrow{\text{linking}} (CH_2O)_n$$

The reduction requires *reducing power* in the form of NADPH and the linking of carbon atoms requires energy in the form of ATP. There is, in fact, no need for the process to go in the order shown: the reduction could come after the linking together. This is simply a very basic and fundamental description of what happens.

The reactions taking place in the chloroplasts are complex, involving many intermediates and many enzymes, but we should not allow this complexity to distract our attention from what is actually happening. For many years it was not known how the process occurred. Then in 1954, Melvin Calvin, in California, started a series of experiments which were eventually to show how the sequence of reactions proceeded.

Calvin exposed unicellular green algae to light in the presence of radio-actively-labelled carbon dioxide ($^{14}CO_2$). He then killed the cells after short periods of time by plunging them into boiling methanol. This stopped all the reactions of photosynthesis instantly. Calvin reasoned that any compounds found in the cells that had become labelled with ^{14}C after such a short exposure should be early products of the 'fixation' of carbon dioxide. ('Fixation' is a rather quaint term meaning to trap an atmospheric gas into chemical compounds: we shall shortly be considering *nitrogen fixation* too.) The algal cells were extracted with solvents and the extracts subjected to paper chromatography (Figure 4.2). The dried chromatogram was placed against photographic film in the dark for a few days, and then developed. Fogging, in the form of dark spots, occurred wherever there was a ^{14}C-labelled compound. Each labelled compound could be identified by its characteristic position on the chromatogram.

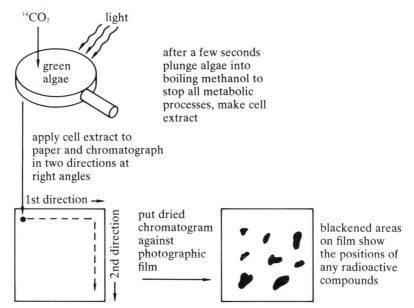

Figure 4.2 Calvin's experiment. The aim of the experiment was to discover the 'first' product resulting from carbon dioxide fixation in the dark reaction of photosynthesis. Green algae were exposed to light in the presence of radioactively-labelled carbon dioxide and the products of photosynthesis analyzed by paper chromatography. Any radioactively-labelled compounds must have been formed from $^{14}CO_2$. Such compounds could be identified by the positions to which they moved on the chromatography paper

Most people at that time believed that the first compound produced in photosynthesis was a C_6 sugar such as glucose. Calvin's experiments showed, surprisingly, that after the very shortest periods of exposure to $^{14}CO_2$, the most heavily-labelled compound was not a C_6 sugar but a C_3 compound, glycerate 3-phosphate. This compound was already known to be an intermediate in the glycolysis pathway. With longer periods of exposure to $^{14}CO_2$ other compounds became labelled, and so it was possible to pick out a sequence of reactions. Painstakingly, Calvin identified all the various intermediates, and eventually was able to show that a cycle of reactions occurred. Calvin was awarded the Nobel Prize in 1961, and the sequence of reactions is now known as 'the Calvin Cycle'. It can be summarized as follows:

(1) Fixation of carbon dioxide. Carbon dioxide reacts with a C_5 compound to yield two mol of a C_3 compound (glycerate 3-phosphate):

$$C_1 + C_5 = C_3 + C_3$$

(2) Reduction and combination. Glycerate 3-phosphate is reduced by NADPH and the product is converted into sugars, amino acids, fatty acids—all the building blocks a plant requires.

(3) Regeneration of the C_5 acceptor molecule. Some of the glycerate 3-phosphate is used in reforming the C_5 molecule so that more carbon dioxide can enter the cycle.

The process is outlined in Figure 4.3: we will now look briefly at the above three processes in more detail.

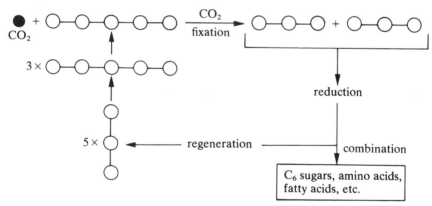

Figure 4.3 Carbon dioxide fixation in the dark reaction of photosynthesis. The net effect of the Calvin cycle is to convert carbon dioxide to organic compounds in plants. Regeneration of the C_5 acceptor molecule (ribulose 1,5-*bis*phosphate) is achieved by converting some of the glyceraldehyde 3-phosphate molecules formed to this compound (five C_3 molecules are converted to three C_5 molecules)

Fixation of carbon dioxide

The key step in the dark reaction of photosynthesis is, of course, the fixing of carbon dioxide into covalent combination in an organic compound. The C_5 acceptor molecule is the diphosphate of a pentose keto-sugar, ribulose (Figure 4.4). The enzyme that catalyzes the fixation is called ribulose *bis*phosphate carboxylase or 'carboxydismutase'. If there is a C_6 intermediate

Overall reaction: $C_1 + C_5 \longrightarrow 2 \times C_3$

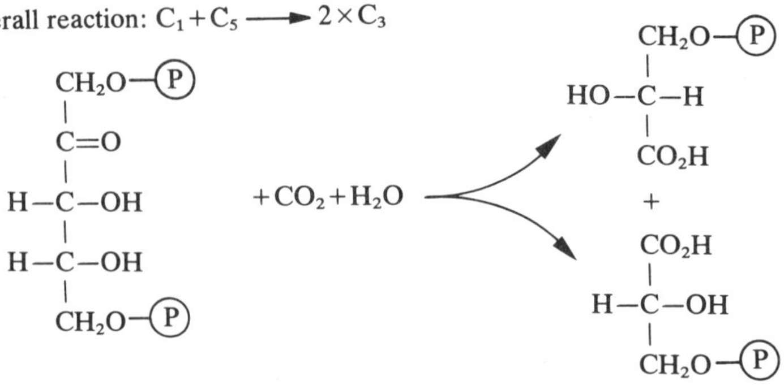

ribulose 1,5-*bis*phosphate

2 mol glycerate 3-phosphate

Figure 4.4 Fixation of carbon dioxide

in this reaction it only exists fleetingly on the surface of the enzyme: the products are 2 molecules of glycerate 3-phosphate. You may notice that neither ATP nor NADPH is used in this process.

The enzyme catalyzing this step of the dark reaction of photosynthesis is present in large quantities in plants (Box 4.1). Some plants that live in regions where the light intensity is high have a supplementary method of fixing carbon dioxide. We shall deal with them later (p. 156).

Box 4.1 The most abundant protein on earth

Ribulose *bis*phosphate carboxylase is the enzyme that catalyzes the carbon dioxide fixing step of photosynthesis. It occurs in the stroma of chloroplasts at a concentration of about 300 mg ml^{-1}. Aqueous extracts of the leaves of higher plants may contain 1–10 mg per g of fresh leaf weight.

Based on the amount of carbon dioxide fixed per year it can be estimated that about 40 000 000 000 000 g of this protein are needed—or about 0.2% of all the protein on earth. It is not certain why this should be. The enzyme seems to be inefficient but it may be that there are chemical constraints that have prevented a 'better' enzyme catalyst evolving. As far as photosynthetic organisms are concerned the solution to the problem of a poor catalyst (for whatever reason) is to have large amounts of it.

Reduction and combination

Glycerate 3-phosphate undergoes two of the reactions of glycolysis, but in reverse. Both ATP and NADPH take part in these reactions and the net effect is to form glyceraldehyde 3-phosphate. In other words a reduction has occurred from a carboxylic acid to an aldehyde:

$$-CO_2H \rightarrow -CHO \qquad \text{(compare Figures 3.34 and 4.5).}$$

This is sometimes referred to as *reductive assimilation* of carbon. In photosynthesis the reduction is done by NADPH, in glycolysis NADH is produced (Figure 4.5). This is typical of most of metabolism: in biosynthetic reactions

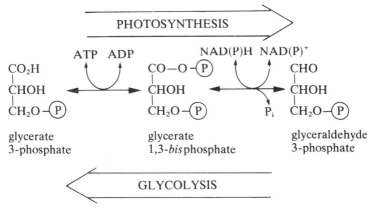

Figure 4.5 Reductive assimilation of carbon dioxide

NADPH participates, but in degradative reactions the use of NAD^+ is the rule.

The empirical formulae of both glucose and glyceraldehyde is $(CH_2O)_n$—in other words the carbon dioxide that has been assimilated now has the same degree of reduction as in carbohydrate (Figure 4.6). In order to convert glyceraldehyde 3-phosphate to glucose, photosynthetic organisms simply use, with one small difference, the reactions of glycolysis in reverse. There is no problem in doing this—enzymes merely catalyze the attainment of equilibrium. However, we might comment that this is yet another example of the economy of pathways in all living organisms. Having once learned a trick, they exploit it for all it is worth!

The one small difference is this: during glycolysis, from glucose to glyceraldehyde 3-phosphate, 2 ATP molecules are used. In the reverse reaction, however, no ATP is generated. Instead, at one step, inorganic phosphate is split off in a reaction catalyzed by a different enzyme from the one operating in glycolysis. Probably, there is not a sufficiently high energy change to produce ATP. Instead, the splitting off of the phosphate group is actually an exergonic reaction which 'pulls' the whole sequence in the direction of carbohydrate biosynthesis:

$$P-C_3 + C_3-P \rightarrow P-C_6-P \xrightarrow[\text{exergonic}]{P_i} C_6-P$$

glyceraldehyde 3-phosphate   glucose 6-phosphate

The glucose 6-phosphate formed could be converted directly into starch, one of the major carbohydrate products of photosynthesis.

Regeneration of ribulose 1,5-bisphosphate

The Calvin cycle is now complete except that the carbon dioxide acceptor must be regenerated to keep the cycle turning. The sequence of enzyme-catalyzed reactions by which this occurs is very complicated, but the principle is easy to understand. Effectively, some of the glyceraldehyde 3-phosphate is 'bled' off, and in a sequence of reactions five C_3-units are converted to three C_5 units. NADPH is not required here, but ATP is. The 'last' compound

$$
\begin{array}{c}
\text{CHO} \\
| \\
\text{H}-\text{C}-\text{OH} \\
| \\
\text{HO}-\text{CH} \\
| \\
\text{H}-\text{C}-\text{OH} \\
| \\
\text{H}-\text{C}-\text{OH} \\
| \\
\text{CH}_2\text{OH}
\end{array}
$$

$$
\begin{array}{c}
\text{CHO} \\
| \\
\text{H}-\text{C}-\text{OH} \\
| \\
\text{CH}_2\text{OH}
\end{array}
$$

$(CH_2O)_3$
glyceraldehyde

$(CH_2O)_6$
glucose

Figure 4.6 Glyceraldehyde has the same oxidation state as glucose

in the sequence is a C_5 phosphate, ribulose 5-phosphate. ATP supplies a phosphate to convert this to the diphosphate, ribulose 1,5-*bis*phosphate, which may be regarded as a highly-activated C_5 sugar ready for carbon dioxide trapping. Although we have not actually seen a reaction in which a carbon–carbon bond is formed using the chemical energy of ATP, this reaction, together with carbon dioxide fixation, effectively achieves the same thing.

C_3, C_4 and CAM plants

It was mentioned earlier that certain plants living in regions of the earth having high light intensities had a 'supplementary' way of fixing carbon dioxide. Although the Calvin cycle essentially, as described above, operates in all green plants, those growing in arid regions in the tropics have a problem. At high light intensities it would be desirable to open up the stomata as wide as possible to allow entry of carbon dioxide for high rates of photosynthesis. Typically, in such regions, the humidity is low, and it is desirable to keep the stomata *closed* to avoid water loss! In order to achieve high rates of photosynthesis without becoming schizophrenic about whether to have the stomata open or closed, certain plants have developed ways of trapping carbon dioxide more effectively at low concentrations, i.e. with the stomata almost closed. They have to 'spend' a little energy in the form of ATP in order to achieve this: presumably this is a worthwhile investment in terms of overall gain. Two ways of doing this are known. A pathway that operates in addition to the Calvin cycle was discovered by M. D. Hatch and C. R. Slack. This is not a minor curiosity, but is of major economic importance as it operates in such plants as sugar-cane and maize.

The Hatch–Slack pathway

'Normal', temperate-zone plants, are called 'C_3 plants' because the first product of the photosynthetic assimilation of carbon dioxide is a C_3 compound, glycerate 3-phosphate. Plants operating the Hatch–Slack pathway, in contrast, are referred to as 'C_4 plants'. Their secret is to trap carbon dioxide at relatively-low concentrations by combining it with phospho*enol*pyruvate (a C_3 compound) to form oxaloacetate (a C_4 compound)—hence the name 'C_4 plants'. Carbon dioxide trapped in this way is released later, elsewhere, to take part in the normal, C_3, Calvin cycle.

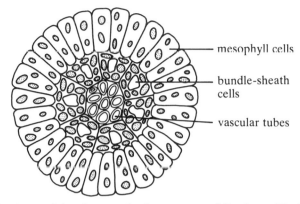

Figure 4.7 Anatomy of the photosynthetic apparatus of C_4 plants. The bundle-sheath cells contain conspicuous chloroplasts

The type of plant operating this 'extra' pathway typically has its photosynthetic cells arranged in bundles around the vascular tubes (phloem) that carry the products of photosynthesis away to other parts of the plant. This is called 'bundle-sheath' tissue (Figure 4.7). The outer mesophyll cells 'spend' ATP to trap carbon dioxide at low concentrations and transport the C_4 product to the inner bundle-sheath cells for use in the Calvin cycle. Both mesophyll and bundle-sheath cells have chloroplasts.

The carbon dioxide-trapping reaction involves the manufacture of the phosphate derivatives of pyruvate at the expense of using an ATP molecule. The phospho*enol*pyruvate is capable of combining with carbon dioxide at low concentration in an enzyme-catalyzed reaction. The sequence is shown in Figure 4.8: if some of the names seem familiar it is because you have already met them in the TCA cycle. The final step is to release the carbon dioxide in the bundle-sheath cells to take part in the Calvin cycle.

CAM plants

Some plants, especially succulents of the family *Crassulaceae* which also live in high light intensity and low water environments, operate a similar mechanism but in a slightly different way. The metabolism of such plants is referred to as Crassulacean Acid Metabolism or CAM. During periods of darkness this type of plant continues to take up carbon dioxide although it cannot be used in photosynthesis. The carbon dioxide is trapped in just the same way as in C_4 plants, except that now the ATP and NADPH are provided by oxidizing previously-formed carbohydrate in the mitochondria, rather than by photosynthesis. The trapped carbon dioxide is stored as acids of the TCA cycle—especially malate, but oxaloacetate and others too—which accumulate in the leaves in the dark. During the next light period the acids are broken down to pyruvate and carbon dioxide, the latter being used in the Calvin cycle.

These ways of fixing carbon dioxide are summarized in Figure 4.9.

The overall result of photosynthesis

The net result of photosynthesis is to provide the plant with glyceraldehyde

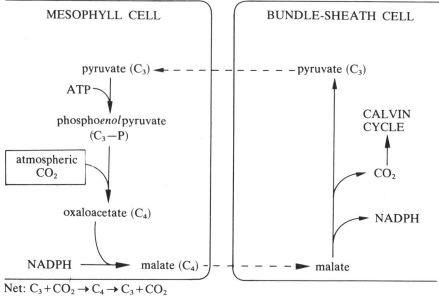

Net: $C_3 + CO_2 \rightarrow C_4 \rightarrow C_3 + CO_2$

Cost: 1 ATP

Regenerated: NADPH

Figure 4.8 C_4 plants—the Hatch–Slack pathway

C_3 PLANTS

$$CO_2 + \text{ribulose-}P_2 \rightarrow 2 \text{ phosphoglycerate}$$

C_4 and CAM PLANTS

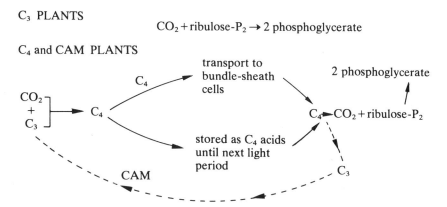

Figure 4.9 Comparison of the different types of photosynthetic carbon dioxide fixation

3-phosphate which may be converted to carbohydrate (glucose 6-phosphate). The energy to drive this synthesis is sunlight trapped in the light reaction of photosynthesis and utilized, as NADPH and ATP, in the dark reaction. Glucose 6-phosphate can be converted to sucrose, a disaccharide, for transport around the plant in the phloem, or alternatively, it may be converted to starch for storage.

Given carbohydrate, plants can synthesize all the other building blocks they need for growth, except that if these building blocks contain nitrogen, this has to be brought in from somewhere. Thus nitrogen is required for the synthesis of amino acids as well as nucleic acid bases. Plants obtain this nitrogen in the form of ammonium ions from the soil or from symbiotic 'nitrogen-fixing' bacteria. This process of nitrogen fixation is of vital importance to all life on earth, and neither plants nor animals are capable of doing it!

4.3 Nitrogen fixation

The earth's atmosphere contains about 79% nitrogen and so represents a great reserve of nitrogen, and yet only prokaryotic micro-organisms have evolved ways of converting it to a form that can be used by plants and other organisms. Micro-organisms fix atmospheric nitrogen by reducing it to ammonium ions. Apart from fixation of atmospheric nitrogen by man in the production of synthetic fertilizers, all life on earth is completely dependent ultimately upon these micro-organisms. In the modern world microbial nitrogen fixation probably accounts for between one-half and three-quarters of the total. One square metre of nodulated legumes with symbiotic nitrogen-fixing organisms can fix between 10 and 30 g nitrogen per year.

Which organisms fix nitrogen?

Some of the micro-organisms that fix nitrogen are free-living in the soil; others live symbiotically in the nodules on the roots of leguminous and other plants (Figure 1.7). *Clostridium pasteurianum*, a bacterium, was shown to fix nitrogen in 1893, but it is now recognized that the ability is widespread amongst prokaryotes, e.g. *Clostridium*, *Azotobacter* and *Klebsiella*. Legumes are by no means the only plants to be 'infected' with nitrogen-fixing micro-organisms. Some angiosperms have symbiotic nitrogen-fixing actinomycetes and some gymnosperms have nitrogen-fixing blue-green algae. Free-living blue-green algae are probably extremely important too in fixing nitrogen. Figures of 2–10 g nitrogen fixed per square metre of water-logged soil have been quoted as attributable to these organisms.

How is nitrogen fixed?

The fixation of nitrogen is a 'difficult' reaction to perform chemically and requires a large input of energy. It can be described by an overall reaction involving the transfer of six electrons:

$$N_2 + 6H^+ + 6e^- \rightarrow 2NH_3$$

(In aqueous systems the ammonia will immediately form ammonium ions.) Although this is the overall reaction, it is not known exactly how the process takes place. It is catalyzed by an enzyme complex called the *nitrogenase complex* which requires a supply of both reducing power and energy. The nitrogenase enzyme complex contains the trace metals iron and molybdenum. This illustrates the vital role of the trace elements.

Ferredoxin supplies the reducing power

The source of the electrons shown in the above overall equation, that is, of the reducing power, is reduced ferredoxin, a compound we have come across previously in photosynthesis (p. 108). Ferredoxin occurs in non-photosynthetic organisms too, and in fact some nitrogen-fixing organisms are photosynthetic and some are not. In the photosynthetic ones, reduced ferredoxin may result directly from photosynthesis. In the non-photosynthetic ones, electrons from fuels such as fats and carbohydrates, obtained by the electron transport chain, are used to reduce ferredoxin. Possibly NADPH is used to reduce ferredoxin in some organisms.

It is noteworthy that in addition to nitrogen, all nitrogen-fixing organisms can also reduce acetylene (ethyne) to ethylene (ethene):

$$HC\equiv CH \xrightarrow{\text{2H}} CH_2=CH_2$$

This forms the basis of the *acetylene reduction test* which allows the nitrogen-fixing potential of organisms to be measured easily.

Cell-free nitrogenase enzyme complexes have been isolated from a number of organisms and all share the property of being readily inactivated by oxygen. This makes them somewhat difficult to work with and certainly impeded early research. Presumably nitrogen fixation takes place in anaerobic regions of the cell.

In addition to the supply of reducing power, ATP is required in large amounts to drive the reaction. Although the number of molecules of ATP required is not certain, it seems that at least 12 are needed for the fixation of one molecule of nitrogen. This may be summed up by another overall reaction:

$$N_2 + 6\,\text{ferredoxin} + {\sim}12\text{ATP} \xrightarrow{\text{nitrogenase}} 2NH_4^+$$
(reduced)
$$+ 6\,\text{ferredoxin} + {\sim}12\text{ADP} + {\sim}12P_i$$
(oxidized)

Nitrifying bacteria

The soil also contains bacteria called nitrifying bacteria, which do not fix nitrogen themselves, but oxidize ammonium ions produced by nitrogen-fixing organisms, and in so doing obtain energy for making ATP for themselves. Nitrate or nitrite may be produced. Many plants benefit from this process since they absorb nitrate ions faster than they do ammonium ions. Plants have the ability to reduce nitrate to ammonium ions which can then be incorporated into amino acids and other nitrogen-containing molecules. The enzyme catalyzing this process is nitrite reductase which also contains molybdenum and iron, and the reducing power used in this case is NADPH. The overall reaction involves nitrite and hydroxylamine (Figure 4.10).

The story so far

By the actions of green plants and micro-organisms, carbon dioxide and atmospheric nitrogen are fixed using reducing power and energy derived from sunlight. The fixed forms of these are converted to all the various organic

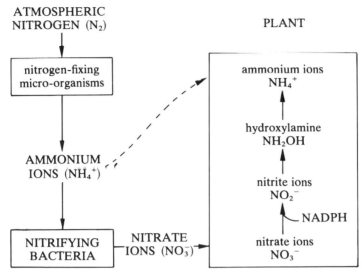

Figure 4.10 Action of nitrifying bacteria. Plants can take up ammonium ions produced by nitrogen-fixing micro-organisms. However, plants absorb nitrate ions much more rapidly than ammonium ions. The nitrate is reconverted to ammonium ion inside the plant and then used to make amino acids and other nitrogen-containing plant constituents

compounds that constitute micro-organisms and plants. Heterotrophs, including animals, use plant and animal tissues as food, so obtaining their building blocks for growth—amino acids, carbohydrates, etc.—in a preformed state. A part of their intake of these compounds is oxidized to obtain energy.

By a whole host of intricate enzyme-catalyzed reactions, all organisms synthesize all the various compounds that constitute their tissues, and the processes are controlled so that the right substances are synthesized at the right times in the right places. The accumulated knowledge about these metabolic pathways fills volumes. We shall merely pause here to look briefly at how storage compounds—starch, glycogen, fats—are manufactured. The study of these pathways will reveal some general rules about biosynthetic pathways, and we shall also note that the processes are practically the same in animals, plants and micro-organisms.

4.4 Biosynthesis for storage

Let us start by reminding ourselves of why it is necessary to store materials. Organisms require a constant supply of energy for continuing their life activities. Yet the sun does not always shine, the herbivore may find no grass, the carnivore may fail to trap its prey. Therefore in times of plenty, stores are put by for lean times (Figure 4.11). The relative advantages of carbohydrate and fat as stores have already been mentioned (p. 127).

Biosynthesis of storage carbohydrate and structural polysaccharides involves linking up monosaccharide units to form a polymer—a condensation reaction. Biosynthesis of fat involves synthesis of long-chain fatty acids from acetyl coenzyme A and then joining these with glycerol to form triglycerides.

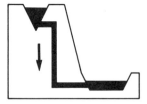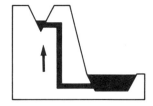

Figure 4.11 Storage. In the heart of a mountain in Wales a pumped storage power station is being built. The power station at Dinorwic is unusual in that it uses water to store energy as potential energy. In times of plenty (i.e. when there is little domestic or industrial demand for electricity) energy is used to pump water up to a high level. The potential energy is tapped by letting the water flow down a 439 m vertical tunnel into a turbine.

In living organisms the potential energy is stored in highly-reduced forms of carbon such as polysaccharides and fats

How are polysaccharides synthesized?

Glycogen is a branched polymer of glucose residues (p. 47) which can be broken down in muscle to supply energy for muscular contraction. A well-fed human has about half a kilogram of glycogen, some in muscle, some in liver.

In Chapter 3 we saw that the enzyme responsible for glycogen breakdown was glycogen phosphorylase:

$$\text{glycogen} + \text{inorganic phosphate} \rightarrow \text{glucose 1-phosphate}$$

and that this enzyme reaction is controlled by the action of the hormone, adrenaline (p. 143). Since enzymes are capable of catalyzing both forward and backward reactions, it is fair to ask: does glycogen phosphorylase catalyze the biosynthesis of glycogen? In fact, the equilibrium constant for the above reaction is about 3.0 and so it *is* possible to drive it backwards in the direction of glycogen biosynthesis by a comparatively modest increase in the *relative* concentration of glucose 1-phosphate. However, this does not seem to happen to any significant extent in living tissue, and it has been discovered that a different set of enzyme-catalyzed reactions is used for synthesizing glycogen! In view of all that has been said previously about the remarkable economy of pathways in living organisms, this would seem to be a bit extravagant!

There is a reason for this apparent duplication of pathways which illustrates a very simple but important principle of metabolism. A liver cell or a muscle cell sometimes needs to synthesize glycogen: at other times it will need to break glycogen down. It is thus necessary to control the direction in which the pathway operates. Control is basically achieved by turning on or turning off enzyme activity. If the same enzyme is used for *both* breakdown *and* biosynthesis of glycogen, switching it off will stop both breakdown and biosynthesis. Switching it on will turn on both breakdown and biosynthesis—no control! The problem is solved by having two slightly different pathways, catalyzed by distinct enzymes. Turning one of these off allows the other pathway to proceed, and *vice versa* (Figure 4.12). Very many metabolic routes, not just those concerned with glycogen biosynthesis and breakdown, are controlled in this way.

ONE PATHWAY, ONE SET OF ENZYMES

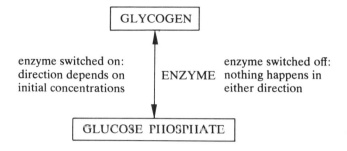

TWO PATHWAYS, TWO SETS OF ENZYMES

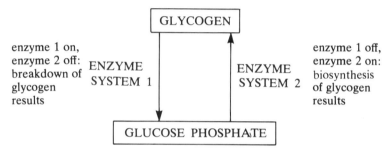

Figure 4.12 Effective control over biosynthesis and breakdown pathways requires two separate pathways. In fact two complete sets of enzymes are unnecessary: it is simply necessary to have a 'loop' at some stage in the pathway where biosynthesis and breakdown pathways differ

The reaction by which glycogen is synthesized is, in fact, only slightly different from the breakdown pathway, but involves the input of energy, which again we should not by now find surprising (see Figure 4.12). In the catabolism of glycogen, glucose units are cleaved in one step. In the biosynthesis of glycogen, two steps are used, one to 'activate' the glucose unit, and another to add it on to the growing glycogen chain. It seems that glycogen molecules shrink or grow as metabolic circumstances demand. Rarely are they completely demolished or completely synthesized from scratch.

The activated form of glucose which can add on to the growing glycogen chain is known by the abbreviation 'UDP-glucose', where UDP stands for *u*ridine *di*-*p*hosphate. Uridine is one of the nucleotides found in nucleic acids (see Chapter 5) and its di- and triphosphates, UDP, and UTP, are analogous to ADP and ATP (Figure 4.13).

UDP-glucose is formed from glucose 1-phosphate in an enzyme-catalyzed reaction with UTP (Figure 4.13). The glucose unit of the UDP-glucose is then added on to a glycogen chain in a reaction catalyzed by the enzyme *glycogen synthase* (Figure 4.14). Note that phosphate bonds of UTP are 'spent' in driving this reaction sequence. In addition, branching enzymes are required to create branch points in the glycogen structure. In this way the biosynthesis and catabolic pathways are made distinct, and individually controllable.

ATP is: UTP is:

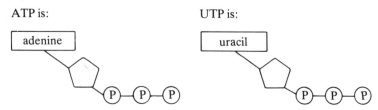

The two compounds differ only in the nitrogen-containing base—either adenine or uracil. Both have a high free energy change for hydrolysis. Reaction between UTP and glucose 1-phosphate, in the presence of a specific enzyme, results in the formation of UDP-glucose. This may be regarded as an *activated* form of glucose

Formation of UDP-glucose

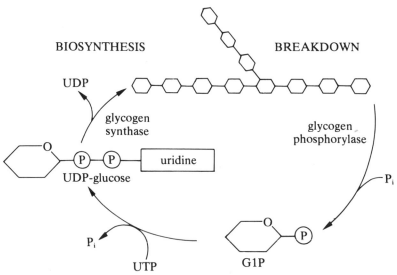

Figure 4.13 Uridine triphosphate and the formation of UDP-glucose

Figure 4.14 The processes of glycogen biosynthesis and glycogen breakdown occur by different pathways with different enzymes. This means that the overall direction of metabolism can be controlled by switching off either the biosynthesis or the breakdown pathway

Another metabolic principle

In glycogen biosynthesis, a *nucleoside diphosphate sugar*, UDP-glucose, is formed and used to add on the sugar unit. When we survey animal, plant and micro-organism metabolism we find that whenever a sugar is added to another sugar, almost always a nucleoside diphosphate sugar participates in the reaction. However, it is not necessarily always UDP-glucose. This general metabolic principle applies not only to polysaccharide biosynthesis but also to the biosynthesis of disaccharides such as lactose (galactosyl–glucose) and sucrose (fructosyl–glucose). It also means, invariably, that biosynthetic and catabolic pathways use different enzymes and that the net direction, bio-synthesis or catabolism, is controllable. Obviously, these synthetic reactions require the input of energy and this is supplied by the hydrolysis initially of a nucleoside triphosphate (see Figure 4.14).

There are many examples of this type of biosynthetic pathway. When starch is formed in plants, *ADP-glucose* supplies the glucose units for the growing starch chain. When sucrose is formed in plants as the major transport form of carbohydrate, *UDP-glucose* donates its glucose residue to fructose to produce the disaccharide (Figure 4.15). Exactly the same sort of mechanisms are involved in the biosynthesis of structural polysaccharides. This illustrates a very useful and important point. When a metabolic pathway has been discovered for one tissue or species, or for one particular compound, the biochemist can often predict with reasonable confidence that similar metabolic routes will be found for other species or compounds. Not only does this demonstrate the unity of biochemistry—it also makes life easier for the biochemist!

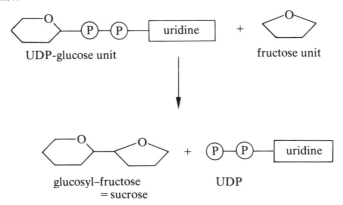

Figure 4.15 The biosynthesis of sucrose also requires UDP-glucose

How is fat manufactured and stored?

As well as storing polysaccharide (glycogen, starch) most organisms also store fat. Fat, and fatty acids, taken in with the diet can be stored, but typically the composition of the fat in food differs from the fat characteristic of the organism that consumes it. Thus the fatty acid chains may be longer or shorter, and there may be greater or fewer double bonds in the molecules. Some dietary fat is deposited unchanged, some is modified and some is synthesized from acetyl CoA units. The relative amounts will depend on the diet and the result will be to form the type of fat characteristic of the particular organism. This is not trivial because the chemical make-up of a fat determines, for example, its

melting point. It would be no use a fish living in arctic waters synthesizing fat
with a composition like that of beef fat—it would set to a solid crystalline mass!

The biosynthesis of neutral fat therefore involves two steps: (1) formation
of fatty acids from acetyl coenzyme A, and (2) linking these with glycerol to
form triglyceride (Figure 4.16). Although the overall reaction is a reversal of
the catabolic pathway, we should not by now be surprised to find that catabolic
and biosynthetic pathways differ in at least some of their reactions so that
the processes can be controlled.

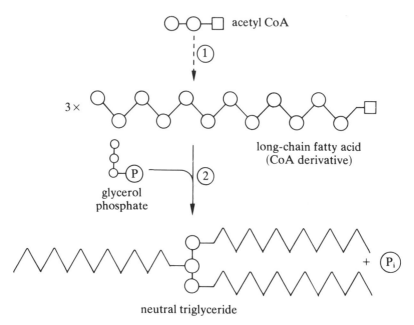

Figure 4.16 The biosynthesis of neutral fat (triglyceride) involves the coenzyme A
derivatives of the long-chain fatty acids as well as glycerol phosphate, rather than
glycerol itself. Note here again that (a) activated forms of the reactants are required,
and (b) the biosynthesis and breakdown pathways differ (compare with glycogen
metabolism)

Biosynthesis of long-chain fatty acids

During the breakdown of fatty acids (p. 126), reduced coenzymes (both NAD^+
and flavins are involved) are produced as a result of the compounds losing
protons and electrons. In synthesizing fatty acids for storage, therefore, a
source of protons and electrons is required. However, as has been said
previously, in most biosynthetic reactions where reduction takes place,
NADPH supplies the reducing power, and this is equally true of fatty acid
biosynthesis. Therefore, the enzymes involved, although catalyzing essentially
the same overall reactions, are different and have a different *cofactor specificity*.
The rate of biosynthesis would, therefore, be dependent upon the NADPH
concentration but independent of NADH and NAD^+ concentrations. In

practice, not only are the enzymes different, but also their location in the cell is different. Fat catabolism takes place largely in the mitochondria. Fat biosynthesis takes place outside the mitochondria, in the cytoplasm, associated with the particulate enzyme complex, fatty acid synthetase.

In addition to the reduction reaction in the process of fatty acid biosynthesis, it is necessary to form carbon–carbon bonds. Therefore, ATP hydrolysis is coupled to the process to supply the driving force.

The reaction sequences catalyzed by the fatty acid synthetase complex are quite complicated, and again differ from the breakdown route. Acetyl coenzyme A combines first with carbon dioxide in a reaction in which ATP is hydrolyzed, not to ADP, but to adenosine monophosphate, AMP, so that effectively, *two* 'high-energy bonds' are used. The product of this reaction is the acetyl coenzyme A derivative of malonic acid, malonyl coenzyme A. This is a C_3 unit, and thus a carbon–carbon bond has been formed.

We might pause here to note that although malonate is a metabolic poison, an inhibitor of the enzyme succinate dehydrogenase (p. 124), malonyl coenzyme A is not, and is actually a metabolite in practically all cells. Probably succinate dehydrogenase does not 'recognize' malonyl coenzyme A as having a complementary shape because of the coenzyme A unit attached to it. Also, malonyl coenzyme A formation is taking place outside the mitochondria whereas succinate dehydrogenase is only found *inside* the mitochondria. In the addition of carbon dioxide to acetyl coenzyme A to form malonyl coenzyme A, a compound called *biotin* participates in the reaction. Biotin is one of the B group vitamins.

The malonyl coenzyme A formed in this first step is highly 'activated' and immediately reacts with another molecule of acetyl coenzyme A. This should form a C_5 compound, but in fact the carbon dioxide that had been added in the previous step, leaves almost simultaneously. The net result is the combination of two C_2 units to form one C_4 unit (Figure 4.17).

The C_4 unit formed is reduced by NADPH, and then the cycle, or rather spiral, starts again, resulting in the formation of a C_6 unit, and so on (Figure 4.17). Eventually, a fatty acid coenzyme A derivative of the correct length for that particular cell type is formed. All of the reactions take place with the various intermediates attached to the fatty acid synthetase complex. This probably speeds up the reaction enormously and makes sure the right chemical operations are performed at the right times.

Formation of triglycerides

In order to form triglyceride, three fatty acid units have to combine with one glycerol unit. This reaction is a condensation one with the formation of three ester linkages. Not only is ATP used, but also the fatty acids will only react if they are in the form of the coenzyme A derivatives. If we remember that it potentially 'costs' ATP molecules to make coenzyme A derivatives, we see that there is quite a high input of energy required to synthesize triglycerides.

Glycerol itself will not react, but rather glycerol phosphate must be formed from glycerol and ATP. Alternatively, intermediates of glycolysis can give rise to glycerol phosphate. The process is summarized in Figure 4.18. The triglycerides formed immediately coalesce to form an inert droplet inside the cell (p. 50).

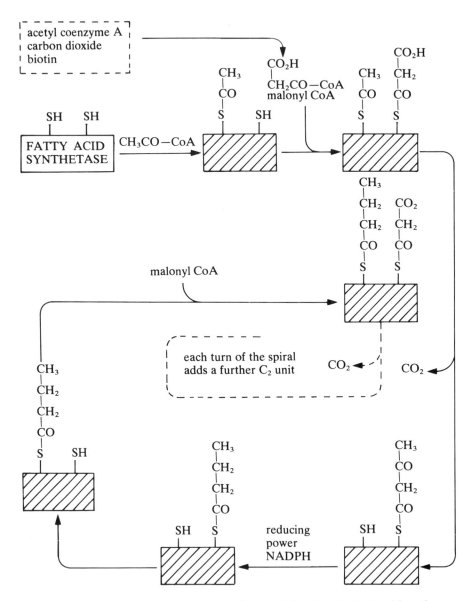

Figure 4.17 The biosynthesis of long-chain fatty acids involves the fatty acid synthetase enzyme complex and both acetyl CoA and malonyl CoA

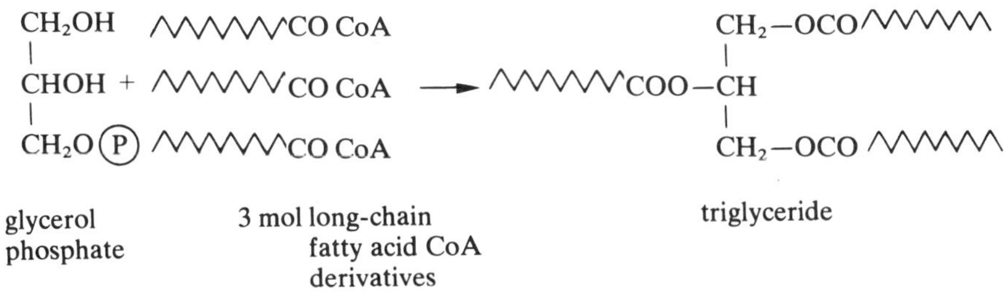

$$\begin{array}{ll}
\text{CH}_2\text{OH} & \text{\Large\char92\char92\char92\char92}\text{CO CoA} \\
| & \\
\text{CHOH} + & \text{\Large\char92\char92\char92\char92}\text{CO CoA} \\
| & \\
\text{CH}_2\text{O}\,\textcircled{P} & \text{\Large\char92\char92\char92\char92}\text{CO CoA}
\end{array}
\longrightarrow
\begin{array}{ll}
\text{\Large\char92\char92\char92\char92}\text{COO} & \text{CH}_2\text{-OCO}\text{\Large\char92\char92\char92} \\
& | \\
\text{-CH} & \\
& | \\
& \text{CH}_2\text{-OCO}\,\text{\Large\char92\char92\char92}
\end{array}$$

glycerol   3 mol long-chain   triglyceride
phosphate   fatty acid CoA
       derivatives

Figure 4.18 Formation of triglyceride for storage requires the coenzyme A derivatives of the long-chain fatty acids and glycerol phosphate

Use of ATP for other life activities

Above, we have seen how ATP hydrolysis is used to drive the biosynthesis of the organic compounds used or needed by living organisms. Sometimes reducing power is needed too, but we should remember that chemical energy in the form of ATP and reducing power in the form of NADPH are really equivalent. Protons and electrons from NADPH *could* be passed down the electron transport chain with the production of ATP from ADP and inorganic phosphate.

ATP supplies the driving force for many other life activities, including the production of light, heat, and electrical and mechanical work (see Figure 4.1). The last of these is of especial interest to animals, and we consider next muscular contraction.

4.5 Muscular contraction

Mobility in biological systems is exemplified by such things as the contraction of animal muscle, the whipping of flagella and the stroking of cilia, the folding of leaves and the movement of the chromosomes during cell division. It is probable that all such processes are driven by free energy obtained during the hydrolysis of ATP, although we have, as yet, a very incomplete understanding of the mechanism of these processes. Mobility driven by ATP is an example of *transduction*, that is the conversion of one form of energy to another (compare Figure 4.1). In the examples mentioned above, the mechanisms involved in producing mobility are probably widely different. However, reduced to the simplest terms, we could say that we start with ATP and an 'extended' structure, and end with a 'contracted' structure and the appearance of ADP and inorganic phosphate. Thus the free energy released in the hydrolysis of ATP is used to produce a change of shape. Although the terms used here are somewhat vague, they should help us to overlook the biological detail and concentrate on the biochemical unity that lies beyond.

Of all forms of biological mobility, the contraction of muscle is the one that has received the most experimental investigation.

What is the source of the ATP used for muscular contraction?

Before we look at the structure and mechanism of action of muscle, let us briefly ask where the ATP used in muscle contraction comes from. Striated muscle can start contracting, at a nervous signal, in a fraction of a second. Hence, consumption of ATP suddenly leaps from a low resting level to a maximum in less than a second, and may continue at this maximal rate for several minutes or longer. It was stressed earlier that ATP is never stored to any appreciable extent, being made on demand by substrate level or oxidative phosphorylation. How can these two apparently opposing notions be reconciled? How can we suddenly increase the ATP supply enormously? It certainly takes time for metabolic pathways to operate and you may remember that the glycolysis pathway actually *consumes* ATP initially before any is produced later. How can this pathway compete for ATP when the muscle is trying to use it up as fast as it can. We can understand the solution to this problem by writing down the sources of ATP supply in voluntary muscle.

Firstly, in the resting state there will be a little ATP around—a few millimol per kilogram of active tissue. However, a man can expend as much as 6 mmol of ATP per kilogram of muscle per second in a sudden burst of activity. Yet the maximum metabolic production of ATP from glycolysis, etc. is about $1 \text{ mmol kg}^{-1} \text{ s}^{-1}$ and it takes several seconds to achieve this rate. The ATP present in the resting state will therefore last for less than 1 second. There must be some other 'immediate' source of ATP, and there is, in the form of a compound called *creatine phosphate.*

Creatine phosphate is one of a group of compounds called *phosphagens.* Creatine phosphate is characteristic of vertebrates, whereas many invertebrates possess phosphoarginine instead. The concentrations of creatine phosphate found in muscle are typically tens of millimoles per kilogram of muscle. Creatine phosphate can react with ADP in a reaction catalyzed by the muscle enzyme creatine kinase, with the formation of ATP and creatine (Figure 4.19). The supply of ATP by this route can therefore last for several seconds— sufficient to fuel a number of vigorous contractions.

Creatine phosphate serves as a buffer-store of instantly-available energy, and during rest periods the reaction shown in Figure 4.19 is reversed until all the creatine is turned into creatine phosphate. (Creatine phosphate is also found in brain tissue, presumably guarding against a temporary lack of ATP.)

The ATP concentration in vigorously contracting muscle does not start to fall for a few seconds because of the 'instant' conversion of creatine phosphate to ATP. After this initial few seconds glycolysis will have had a chance to start up (p. 129) so releasing energy anaerobically from glycogen. Finally, contractions either stop or become less strenuous and more rhythmic, so that the muscle eventually slips into the aerobic mode of energy supply, using either carbohydrate or fat as fuel.

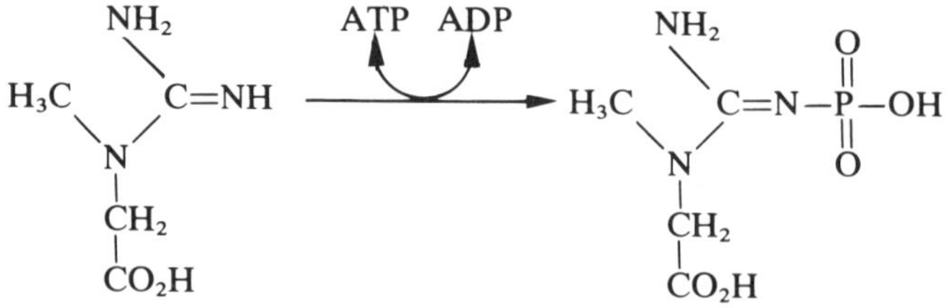

Figure 4.19 Creatine and creatine phosphate

We now return to the question of how the potential chemical energy of hydrolysis of ATP is turned into mechanical energy.

Structure of striated muscle

In skeletal muscle of vertebrates, as well as in the flight muscles of insects, the cells of the muscle tissue fuse to form fibres. These may be 50–100 μm in diameter and several centimetres long. The fibres can shorten in length, and this is how motion is generated. Connective tissue joins the ends of fibres, eventually, to the skeleton so that when contraction occurs limbs are moved, and so on (see Figure 4.20).

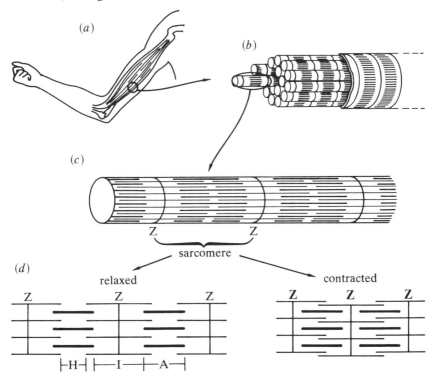

Figure 4.20 Vertebrate striated muscle. (*a*) The muscle tissue is made up of fibres. (*b*) One fibre, (*c*) one myofibril. (*d*) The A band corresponds to the thick myosin filaments and the lighter H zone is where *only* thick filaments occur. On either side of this H region but still within the A band are darker regions where both thick (myosin) and thin (actin) filaments overlap. Thin filaments *only* make up the I regions and the Z line is the structure to which the thin filaments are joined

The fibres themselves are really multinucleate cells. They are packed with smaller fibrous structures called *myofibrils* with the nuclei and mitochondria laying on the outside. It is the myofibrils that shorten during contraction. Each myofibril is composed in turn of units called *sarcomeres* and the repetition of these units gives the muscle its characteristic striated pattern. The dark lines between sarcomeres are called Z-lines (Figure 4.21). Each sarcomere is

(a)

(b)

Figure 4.21 Muscle structure. (a) Shows an electron micrograph of a section of a piece of mouse back muscle cut parallel to the direction in which the fibres run. Note the dark Z lines and also the glycogen granules. (*Photograph kindly supplied by Douglas Kershaw, University of Leeds*) (b) Shows a section of a piece of Tsetse fly muscle cut at right angles to the direction in which the fibres run. The dark areas between the muscle fibres are mitochondria (see also Fig. 3.19). (*Photograph kindly supplied by Dr M. Anderson, University of Birmingham*) Compare with the diagrammatic interpretations of those in Fig. 4.20 (c) and (d). In both pictures, length of line 1 μm

composed of a set of protein filaments, thick and thin, arranged longitudinally. Their arrangement can be seen in electron micrographs where the muscle has been cut at right angles to their direction. These electron micrographs have been interpreted to indicate that the thick and thin filaments 'interdigitate'. The Z-line is a structure to which thin filaments are anchored at their mid-points. The I region (see Figure 4.20) is where there are *only* thin filaments, and the H region is where there are *both* thick and thin filaments.

It is believed that the contractile unit, a sarcomere, can shorten by the fibres sliding between one another. In other words, the zone of overlap between thick and thin filaments increases until the thin filaments meet and actually overlap slightly. The sliding thus pulls the Z-lines closer together and the fibre shortens. This theory accounts for the changes observed   but is it possible to explain in chemical terms how ATP can bring about the sliding and hence contraction?

Actin and myosin

The thick filaments are composed of a protein called *myosin*, and the thin filaments are composed of another protein called *actin*. Each myosin molecule is thought to be made up from two intertwined polypeptide chains, with the end of each chain folded into a globular structure. A myosin filament (thick filament) is made up of a number of these molecules lying side by side (Figure 4.22). The (thin) actin filaments are attached to plates (Z-bands), and the actin molecules are made up of two intertwined chains each of which is rather like a string of beads.

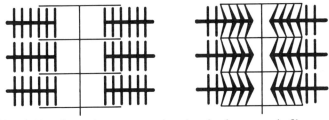

Figure 4.22 Highly schematic representation showing how myosin fibres move against actin filaments

It is proposed that cross-bridges form between the globular heads of the myosin molecules and the beads of the actin molecules. Indeed cross-bridges can be seen in electron micrographs. In order to achieve the sliding action it is necessary to envisage the cross-bridges as being moveable. This could occur if the globule of the myosin molecule could move relative to the shaft of the molecule, and such a conformational change could be driven by ATP hydrolysis. Since the two ends of the thick filaments are effectively pulling in opposite directions on the thin filaments the result is that the two sets of thin filaments are pulled towards each other. Backward slippage does not occur because the thick filament has many myosin heads and not all of them are in the same part of the moving cross-bridge cycle at the same time.

It is known that myosin shows ATPase activity, that is it catalyzes the hydrolysis of ATP to ADP and inorganic phosphate, and it has been shown that the active site responsible for this activity is in the globular head of the molecule. However, little is known of how the conformational change occurs, and much indeed remains to be discovered about how muscles work. In addition to the two proteins actin and myosin, and ATP, other proteins are involved, as well as magnesium and calcium ions. Calcium ions are involved in the triggering of a contraction by a nervous impulse.

Triggering contractions

The myofibrils are set in an extensive network of tubules called the *sarcoplasmic reticulum* (Figure 4.23). The arrival of a nerve impulse to the muscle causes a discharge of stored calcium ions from the sarcoplasmic reticulum into the cytoplasm of the fibre. This release of calcium somehow facilitates the enzyme-catalyzed splitting of ATP to ADP and eventually the interaction between actin and myosin. Hence, shortening of the sarcomere takes place.

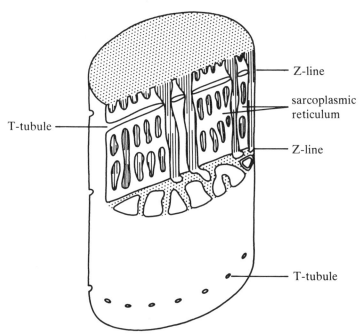

Figure 4.23 Cutaway drawing of part of a muscle fibre showing the intimate connections between the sarcoplasmic reticulum, the T-system and the muscle fibres. The sarcoplasmic reticulum does not open to the outside, but the T (or transverse) system does. There is no direct connection between the two systems. These systems function in triggering the muscle to contract when a nerve impulse is received

Cilia and flagella

Eukaryotic cells often possess cilia and flagella which have the property of movement. Cilia keep a film of liquid moving parallel to the cell surface by beating in a co-ordinated fashion. Flagella propel protozoa and also sperm cells. Is this movement driven by ATP? The answer is almost certainly 'yes', and the movement results from an action very much akin to that occurring in voluntary muscle.

When cilia and flagella are observed in section in the electron microscope the same fundamental design is nearly always observed (Figure 4.24). Down the centre of each cilium or flagellum runs a bundle of fibres called an axoneme which is composed of a set of two *microtubules* surrounded by an outer ring of nine microtubules, these outer ones being double tubes or doublets. Associated with these microtubules is a protein having ATPase activity called *dynein*. It seems that the outer doublet microtubules can slide past each other to produce a bending motion. Dynein molecules joined to one doublet microtubule can 'walk'—that is form a sequence of cross-bridges—along the surface of the adjacent doublet, as ATP is hydrolyzed. This is exactly analogous to the way in which myosin forms cross-bridges with an actin filament in skeletal muscle.

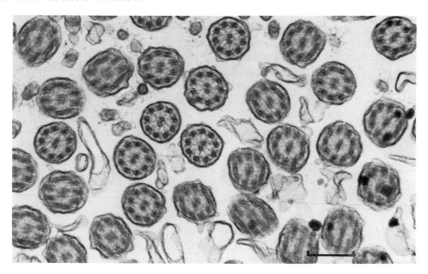

Figure 4.24 Cilia in rat tracheal epithelium seen in transverse section in the electron microscope. Each cilium (like the flagella of eukaryote cells such as sperms) has a highly characteristic 9 + 2 arrangement of the internal 'microtubules'. A protein called dynein in cilia and flagella, has ATPase activity and it is believed to undergo conformational changes which allow the microtubules to 'walk along' or slide against one another bringing about bending of the ciliary stalk. Length of line 0.25 μm

Bacterial flagella

Motile bacteria swim by rotating flagella, rather than by beating them. Bacterial flagella are thin filaments made of a protein called *flagellin*, and their structures are very much simpler and smaller than eukaryotic flagella. For example, a cell of *Escherichia coli*, a common gut bacterium, typically has six flagella which are about 15 nm in diameter and up to 10 μm long.

Bacterial flagella do not actively bend because they have no contractile element. Rather the whole structure appears to be rotated by a 'motor' associated with the basal structure of the flagella. The rotation of bacterial flagella is not driven by the hydrolysis of ATP! Instead it is driven by a proton-motive force, effectively a proton gradient, across the bacterial plasma membrane. However, we know that these two—a proton gradient and ATP—are really equivalent, since one can be converted to the other in oxidative phosphorylation or photosynthesis.

In bacteria, the velocity of rotation of the flagella is directly proportional to the proton-motive force. In one second an *E. coli* cell can swim about fifteen times its own length—a distance of about 30 μm. (This is equivalent to the present world record for the 100 m sprint, or to about five times the present record for swimming 100 m.)

Questions

1. In an experimental investigation of the incorporation of carbon dioxide into photosynthesizing cells a suspension of unicellular green algae was placed in a container giving optimum conditions for photosynthesis. The suspension was stirred mechanically and radioactive carbon dioxide was added in the form of a solution of $NaH^{14}CO_3$. At regular intervals samples of the suspension were removed, plunged into hot alcohol and then analyzed by two-dimensional chromatography. After development of the chromatograms the individual compounds separated were identified and the amount of radioactivity in each was measured. Percentages of total radioactivity calculated from a typical set of results are given in the table.

Time after addition of $NaH^{14}CO_3$ solution (seconds)	Percentages of total radioactivity in:			
	glucose phosphate	pentose phosphate	phosphoglyceric acid	other organic compounds
20	67	6	27	0
40	60	27	4	9
80	50	33	0	17
140	28	46	2	24

It can be assumed that the non-radioactive atoms of ^{12}C are fixed at the same rate as the atoms of ^{14}C.

(*a*) With a single set of axes plot graphs of the percentages of total radioactivity present in the three identified compounds against time.

(*b*) Examine the curves produced and state what conclusions may legitimately be drawn from them.

(*c*) Suggest why unicellular algae were used as experimental material.

(*d*) Explain why the suspension was stirred.

(*e*) Suggest why the samples of suspension removed were plunged into hot alcohol.

(*f*) Suggest a simple modification of experimental procedure which would permit the removal of the first sample sooner than 20 seconds after the introduction of the radioactive material.

(*g*) Explain what sort of information you think would be obtained with a sample removed 5 seconds after the introduction of radioactive material.

[AEB, 1973]

2. Give an account of the fixation and utilization of carbon in plants which is sufficiently detailed to explain the following statements.

(*a*) (i) The first product of carbon dioxide fixation is 3-phosphoglyceric acid (PGA).

(ii) Light energy is required only to provide the ATP and electron donors (reduced coenzymes) needed to reduce PGA and maintain a supply of ribulose diphosphate.

(iii) There are alternative uses for ATP and the electron donors in addition to those of carbon dioxide fixation.

(b) (i) Starch grains begin to grow close to grana.

(ii) The formation of cellulose and pectins occurs near to the plasmalemma.

[JMB]

3. (a) Name *one* example of an economically-important C_3 plant and *one* example of an economically-important C_4 plant.

(b) What are the advantages to a plant to be capable of the C_4 mechanism of carbon dioxide fixation.

(c) Describe briefly the typical arrangement of tissues in a C_4 plant.

4. State: (a) which of the following require an input of metabolic energy (i.e. must be coupled to reactions involving ATP), and (b) which require an input of 'reducing power'.

(i) Conversion of glucose to fructose 1,6-*bis*phosphate.

(ii) Conversion of fatty acids to acetyl coenzyme A.

(iii) Conversion of glycogen to glucose 1-phosphate.

(iv) Fixation of nitrogen ($N_2 \rightarrow NH_4^+$).

(v) Conversion of glycerate 3-phosphate to pyruvate.

(vi) Fixation of carbon dioxide in photosynthesis.

(vii) Passage of electrons from cytochrome c to oxygen.

5. Briefly explain the following:

(a) chemical composition of fats,

(b) digestion of fats in mammals,

(c) uptake of the products of fat digestion from the alimentary canal in mammals,

(d) storage of fats,

(e) importance of phospholipids.

[O & C]

6. Discuss the following statements regarding ATP.

(a) All metabolizing cells require a constant supply of ATP.

(b) ATP is generated and used up in the process of photosynthesis.

(c) ATP is essential in the functioning of muscles.

[O & C]

7. (a) Only a small fraction of the light falling on a leaf is actually absorbed and used in photosynthesis. State *two* possible fates for the light which is not absorbed.

(b) State precisely where, within a chloroplast, you would find molecules of chlorophyll.

(c) In working out the mechanism of the dark reaction of photosynthesis, the unicellular alga, *Chlorella*, was used. Bearing in mind the techniques that were used, explain the advantages of using an organism of this type.

(d) What is the advantage of using an isotope of an element when studying a complex chemical process such as photosynthesis?

(e) Name *two* isotopes that were used to study the mechanism of photosynthesis and explain briefly what information was derived from their use.

(f) Complete the following sequence of reactions by inserting the following three substances in the appropriate space in the diagram overleaf.

RUDP—(ribulose 1,5-*bis*phosphate, a five-carbon compound)

PGA—(phosphoglyceric acid, a three-carbon compound)

glucose

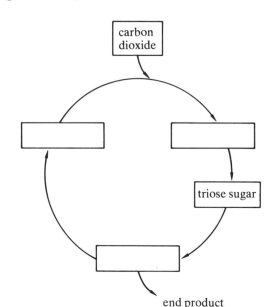

end product

(g) What name is given to the sequence of events shown in the previous question? (v) Krebs' cycle, (w) Hill reaction, (x) citric acid cycle, (y) Calvin cycle, (z) photolysis.   [Scottish Higher]

Note: glycerate 3-phosphate is referred to in some of these questions as 'phosphoglyceric acid', '3-phosphoglyceric acid', and 'PGA'.

Further reading

Bjorkman, O., and Berry, J. (October 1973), 'High Efficiency Photosynthesis', *Scientific American*, **229**, 80.

Brill, W. J. (March 1977), 'Biological Nitrogen Fixation', *Scientific American*, **236**, 68.

Buller, A. J. (1975), 'The Contractile Behaviour of Mammalian Skeletal Muscle', *Oxford/Carolina Biology Reader*, **36**.

Hall, D. O., and Rao, K. K. (1977), 'Photosynthesis', *Studies in Biology*, **37**, (Second Edition). (Edward Arnold).

Harrington, W. F. (1980), 'Theories of Muscular Contraction', *Oxford/Carolina Biology Reader*, **114**.

Huxley, H. E. (December 1965), 'The Mechanism of Muscular Contraction', *Scientific American*, **213**, 18.

Margaria, R. (March 1972), 'The Sources of Muscular Energy', *Scientific American*, **226**, 85.

Moore, P. (February 1981), 'The Varied Ways Plants Trap the Sun', *New Scientist*, p. 394.

Murray, J. M., and Weber, A. (February 1974), 'The Cooperative Action of Muscle Proteins', *Scientific American*, **230**, 58.

Saffrany, D. R. (October 1974), 'Nitrogen Fixation', *Scientific American*, **231**, 64.

Satir, (October 1974), 'How Cilia Move', *Scientific American*, **231**, 44.

Walker, D. (1979), 'Energy, Plants and Man'. (Packard Publishing).

Whittingham, C. P. (1971), 'Photosynthesis', *Oxford/Carolina Biology Reader*, **9**, (Second Edition).

Wilkie, D. R. (1976), 'Muscle', *Studies in Biology*, **11**, (Second Edition). (Edward Arnold).

5

Biochemistry and Genetics

Summary

The ability of living cells to reproduce themselves requires that they have a store of information telling them how to construct their own characteristic types of macromolecules. This information must be duplicated when cells divide. Much of the information represents plans showing how to build enzymes and other proteins, and is stored in the DNA in the form of a code as a sequence of nucleotide bases along the molecule. Translating this information into protein is a complicated process requiring the participation of RNA and subcellular organelles called ribosomes, as well as the input of energy in the form of ATP. Any change in the coded message stored in the DNA may lead to the alteration of an amino acid sequence in a polypeptide, which is called a mutation. Although the DNA contains the information for all the proteins a cell can possibly make, not all are, in fact, made at a given time. There is control of gene expression therefore. The mechanisms whereby this is achieved in prokaryotes are quite well understood, but the mechanisms operating in higher organisms (which have very many more genes) are only just beginning to be elucidated.

5.1 Introduction

Living organisms can duplicate themselves. Seen at its very simplest, a cell has the ability to divide to give two apparently identical daughter cells. How does this happen? What is the chemical explanation for this fundamental ability of living organisms? Can each type of molecule within the cell duplicate itself? If so, it is not a type of chemical reaction we have come across so far.

Since all the processes that go on in a cell are controlled by enzymes, it is easy to see that provided enzyme proteins can be made, then all the rest of metabolism will proceed. Therefore, in this chapter we ask how information specifying the order in which to join amino acids together to make a protein can be (a) stored, and (b) used when new cells form.

In the past it was thought that the only molecules that were of sufficient complexity to carry this sort of information were the proteins themselves. This led to a circular argument. If proteins specify enzyme proteins, what specifies the proteins that specify the enzyme proteins, and so on! Over the years biochemists realized that it was the class of molecules called nucleic

179

acids, and especially the deoxyribonucleic acid, or DNA, of the nucleus that carried the blueprints for making proteins. To be precise these molecules carry, in coded form, the information specifying the sequence in which to link amino acids to form proteins. The details of how this system works are very complex, but the basic ideas are very simple.

A major leap forward was taken in 1953 when the structure of DNA was established. Since then there has been an explosion of knowledge in this area of biochemistry, referred to as 'molecular biology'. This knowledge provides explanations for a whole host of biological phenomena: mutations, virus infections, inherited diseases, and many more. Our starting point is the area of biology dealing with the inheritance of characteristics, that is *genetics*. We shall finish by being able to describe 'the chemical nature of the gene' and a great deal more!

5.2 Biochemistry and genetics

It has been known from time immemorial that offspring inherited some of their parents' characteristics. Mendel started the scientific study of this inheritance. Later it was established that, in eukaryotes, the so-called genes were carried on the chromosomes of the nucleus. The genes were thought of as the physical units of inheritance, but although it was shown that they were arranged in a linear sequence on the chromosomes, their chemical nature was a mystery. They were thought of simply as strings of beads that in some miraculous way controlled all the characteristics of living organisms. It was a major advance to establish that the 'genetic material' forming the genes was in fact DNA.

What is the nature of the genetic material?

Biochemists began to answer this question by analyzing the composition of the chromosomes. Miescher, in 1869, isolated nuclei from the white blood cells found in pus. He discovered that the proteolytic enzyme pepsin did not 'digest' the isolated nuclei, although it caused them to shrink somewhat. The nucleus therefore contained something other than protein. He isolated a non-protein, phosphorus-containing material from the nuclei which he called 'nuclein'. As it was strongly acidic it was, some 20 years later, renamed 'nucleic acid'. Miescher's nucleic acid was found in white blood cells and sperm cells, and thymus gland was also a very good source. When, shortly, another nucleic acid differing slightly in properties was isolated from yeast, it was called 'yeast nucleic acid' in order to distinguish it from Miescher's 'thymus nucleic acid'. Thymus nucleic acid is DNA and yeast nucleic acid is RNA, or ribonucleic acid, and for a time it was suspected that animals had only DNA and plants only RNA. By the 1940s, however, it was realized that both 'acids' were present in both animals and plants.

The important conclusions resulting from this early research was that the nuclei of all cells contained these phosphorus-containing acids. The connection between this and inheritance of characteristics emerged when it was shown by histological staining methods that the DNA of the nucleus was restricted to the chromosomes. It was already known that chromosomes, like genes, came in pairs, one of which was inherited from the father and one from the mother. Soon it was established, using the staining methods, that all the somatic cells of a given organism contained the same amount of DNA (Table 5.1). The amounts of other substances in the cell varied widely. For example, an adipose tissue cell contains a lot of fat but little glycogen while a liver cell would contain a considerable amount of glycogen but little fat. The amount of DNA in each, however, was exactly the same. Furthermore, when the gametes (sperms, egg cells) were examined, they were found to have exactly *half* this amount of DNA (and, of course, they have one set of chromosomes instead of two). This fitted in nicely because at fertilization two gametes would combine, reconstituting the full complement both of DNA and of chromosomes.

Table 5.1 DNA content of cells

	DNA per cell (pg*)
Bacteria	approx. 0.05
Fungi	approx. 0.10
Green plants	2.5
Molluscs	1.2
Fish	2.0
Mammals	6.0
Chicken heart	2.45
Chicken liver	2.66
Chicken pancreas	2.61
Chicken sperm	1.24

* pg = picogram (1 pg = 1×10^{-12} g)

Conclusions
(1) The more complex the organism (i.e. the more different kinds of protein that it can synthesize) the greater the content of DNA per cell.
(2) All the (diploid) cells of a given organism have the same amount of DNA.
(3) The haploid cells (sperms, eggs) have almost exactly half this amount.

This strongly suggested that DNA was the genetic material. Many workers were sceptical, however. They knew that the chromosomes contained both DNA *and* protein, and from what was then known of the structure of DNA, nucleic acids were thought to be far too simple to carry the genetic instructions for a whole organism. The evidence for DNA as the genetic material was considered to be only circumstantial: direct proof was lacking.

Direct evidence that DNA is the genetic material

The structure of DNA will be described later. First, it is necessary to understand how it was established that DNA is indeed the genetic material. Surprisingly, perhaps, we have to turn away from organisms that possess well-defined chromosomes to those which do not, namely the prokaryotes, to find direct proof.

The first real proof came from some experiments performed with the bacteria that cause pneumonia, the pneumococci. In 1928 Griffith, a bacteriologist, reported some curious research findings with these organisms. Certain strains of pneumococci have an extracellular polysaccharide layer or capsule. These strains form smooth, glutinous colonies on agar plates, and are called 'S' strains—S for smooth. The same strains are also virulent, that is they cause pneumonia and death in mammals such as mice. Griffith found that if 'S' cells were heat-killed by boiling, and then injected into mice, the mice survived because the dead cells no longer multiplied inside the mouse. In contrast to these virulent types of cell, other strains of pneumococci were known which had no polysaccharide capsule and which did not cause pneumonia when injected into mice. These are non-virulent or 'R' cells; 'R' because they formed rough-looking colonies on agar plates. Both 'S' and 'R' strains bred true: the ability to produce polysaccharide capsule was genetically determined. It was not possessed by the 'R' cells.

The curious finding of Griffith was that if he took a mixture of heat-killed 'S' cells and live 'R' cells—neither of which should be fatal—and injected it into mice, the mice contracted pneumonia and died (Figure 5.1)! He examined the dead bodies and found large numbers of *live* 'S' cells which he had certainly not put there! Apparently, live 'R' cells had been 'transformed' into 'S' cells by the presence of dead 'S' cells. It seemed that a hereditary property, the ability to produce a polysaccharide capsule, had been transferred from dead to live cells. Within a couple of years this experiment had been repeated in the test tube using a cell-free extract of dead 'S' cells and live 'R' cells. However, the full significance of this was not really appreciated for another 10 years. The key experiment was to mix live 'R' cells with absolutely pure DNA from 'S' cells and to show that this resulted in transference of the ability to produce polysaccharide capsule to the 'R' cells. Hence, the 'information' for producing polysaccharide capsule resides in the DNA and not in protein.

The Hershey and Chase experiment

Some were still sceptical about the ability of DNA to carry information, but evidence continued to accumulate. An experiment was performed by Hershey and Chase and published in 1952, using the viruses called bacteriophages that infect bacterial cells. Their experiment was important because it not only showed that DNA was the genetic material, but also showed how viruses invade cells. The experiment was very ingenious in the way that radioactive isotopes were used.

The bacteriophages (usually known as 'phages' for short) employed by Hershey and Chase were those belonging to the class called 'T-phages'. These are quite complex structures as viruses go and have a 'head' made up of one very long molecule of DNA coiled up inside a protein 'coat', and a 'tail' which is a narrow, protein tube (Figure 5.2). When infection takes place the 'tail' sticks to the cell wall of the bacterium, a special enzyme of the T-phage makes a hole, and the virus DNA molecule is 'injected' into the bacterial cell. Very soon after this has happened the economy of the bacterial cell is altered to making new T-phage particles. Eventually, the cell bursts with the release of several hundred new bacteriophages. Thus when infection takes place, the genetic information for making bacteriophage DNA and protein passes into the bacterial cell. The Hershey and Chase experiment showed that the only thing transferred from bacteriophage to bacterium was DNA.

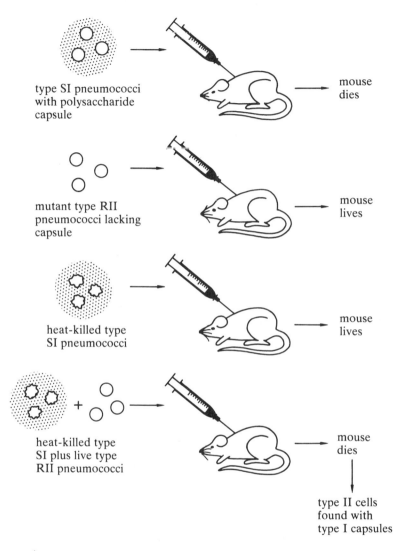

type SI pneumococci
with polysaccharide
capsule

mouse
dies

mutant type RII
pneumococci lacking
capsule

mouse
lives

heat-killed type
SI pneumococci

mouse
lives

heat-killed type
SI plus live type
RII pneumococci

mouse
dies

type II cells
found with
type I capsules

Figure 5.1 Griffith's experiment
Only pneumococcus bacteria having an extracellular polysaccharide capsule cause
pneumonia and death in mice. The ability to produce a polysaccharide capsule is
genetically determined. Polysaccharide-producing strains are called 'S' strains because
the colonies they form on agar plates are 'smooth'. Strains lacking a polysaccharide
capsule are non-virulent, and are designated 'R' strains because they produce colonies
on agar plate that have a 'rough' appearance. The Griffith experiment is shown above.
The conclusion from the last part of the experiment was that the genetic message for
making polysaccharide capsule had somehow passed from dead 'S' cells to live 'R'
cells, giving them the ability to make a capsule. This process is called *transformation*.
It was later found that if 'R' cells were mixed with absolutely pure DNA from 'S'
cells the same result was obtained. This showed that of all the compounds present in
the 'S' cell it was the DNA that carried the genetic information

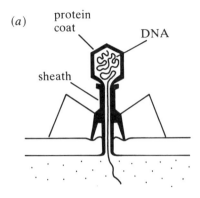

Figure 5.2 (*a*) Bacterial viruses or bacteriophages of the type called 'T phages' are composed of a single double-helical length of DNA inside a protein coat. There is a 'tail' in the form of a hollow protein tube, surrounded by a sheath, with tail fibres which attach to the cell wall of the bacterium. (*b*) In this shadowed preparation seen in the electron microscope a single bacteriophage has been caught in the act of injecting its DNA into a bacterial cell. Length of line 0.1 μm

They did their experiment by making use of some chemical knowledge. Bacteriophage proteins contain sulphur in the amino acids cysteine and methionine (see Table 2.3), but do not contain phosphorus. Bacteriophage DNA contains phosphorus but no sulphur. Therefore propagating bacteriophage on bacteria growing in a medium containing radioactive sulphur (^{35}S) will yield bacteriophage particles labelled in the protein *only*. In contrast, using medium containing radioactively labelled phosphate (^{32}P) will label the DNA *only*. Hershey and Chase grew and isolated both 'varieties'

(*a*) Infection

(*b*) 'Injection'
of DNA

(*c*) Protein of virus
stays on the outside:
DNA (and genetic
information) passes
into the cell

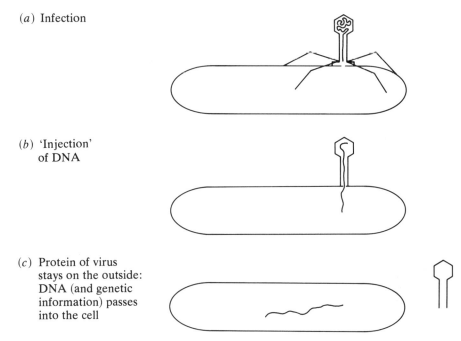

Figure 5.3 The Hershey and Chase experiment. When a bacterial virus (bacteriophage) infects a bacterial cell, the protein coat of the virus stays outside the cell and 'injects' its DNA into the cell. Thereafter the cell contains viral genetic information, i.e. the information with which to make virus-type proteins. Hershey and Chase showed that this was the case by labelling the protein coat of the virus with ^{35}S—in which case all the radioactivity was found outside the bacterial cells, or by labelling the DNA with ^{32}P, in which case all the radioactivity was found inside the cells

of bacteriophage, protein-labelled and DNA-labelled. They found they could mix bacteria and bacteriophage, allow infection to take place for a few minutes, and then separate them again by violent agitation. What was left of the bacteriophage could then be separated from the bacterial cells by centrifugation. When the experiment was done with DNA-labelled bacteriophage all the radioactivity was found to pass into the bacterial cells (Figure 5.3).

Such experiments demonstrated that without doubt DNA carried the 'genetic message'. We cannot proceed much further without an understanding of the chemical structure of DNA.

5.3 The structure of DNA

It has long been known that the macromolecule, DNA, is made up from three components: nitrogen-containing bases, carbohydrate (deoxyribose) and phosphate. There are 4 types of nitrogen-containing base in DNA. They are adenine, guanine, cytosine and thymine (Figure 5.4). The first two have the double-ring *purine* structure: the second two have the single-ring *pyrimidine* structure. The sugar, deoxyribose (Figure 5.5), forms, with phosphate groups, a very long backbone, alternating sugar–phosphate–sugar–phosphate, and so

purine nucleus 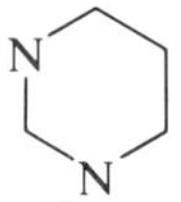   pyrimidine nucleus

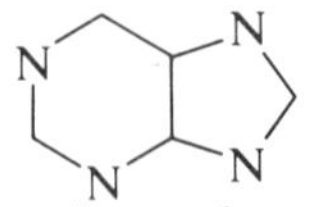

The structures of the four bases are as follows: the arrows show where the nitrogen-containing ring is joined on to the backbone (see Fig. 5.5)

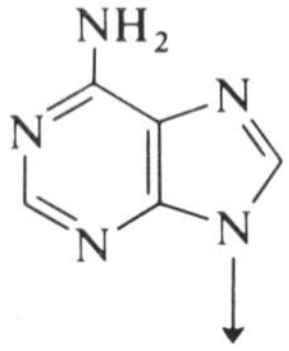

adenine (A) 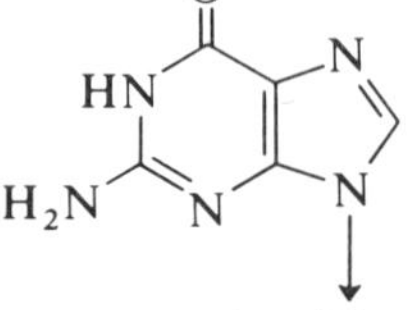   guanine (G) 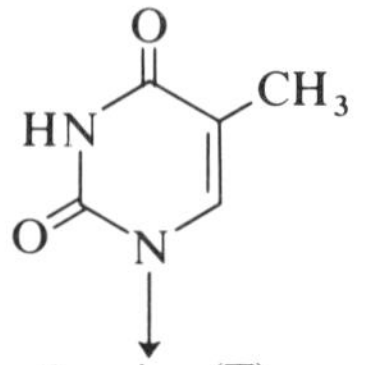   thymine (T) 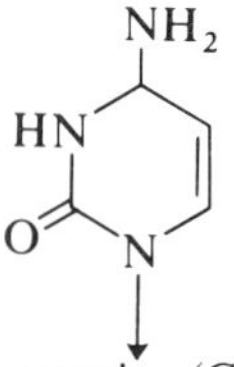   cytosine (C)

Figure 5.4 Structure of DNA. DNA from whatever source contains the four nitrogen-containing bases adenine, guanine, thymine and cytosine. The first two of these are built up from the double ring *purine* nucleus, and the second two are built up from the single ring *pyrimidine* nucleus.

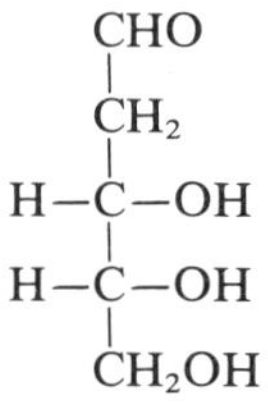

$$CHO$$
$$CH_2$$
$$H-C-OH$$
$$H-C-OH$$
$$CH_2OH$$

2-deoxyribose

$$CHO$$
$$H-C-OH$$
$$H-C-OH$$
$$H-C-OH$$
$$CH_2OH$$

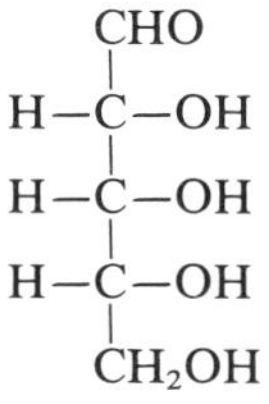

ribose

Both DNA and RNA have a repeating sugar–phosphate backbone:

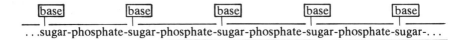

...sugar-phosphate-sugar-phosphate-sugar-phosphate-sugar-phosphate-sugar-...

As shown, one of the four bases is attached to each of the sugar units. Nucleic acids can therefore be thought of as strings of base–sugar–phosphate units: such a unit is called a nucleotide. (Confusingly a base—sugar unit is called a nucleoside: ATP is therefore a nucleoside triphosphate)

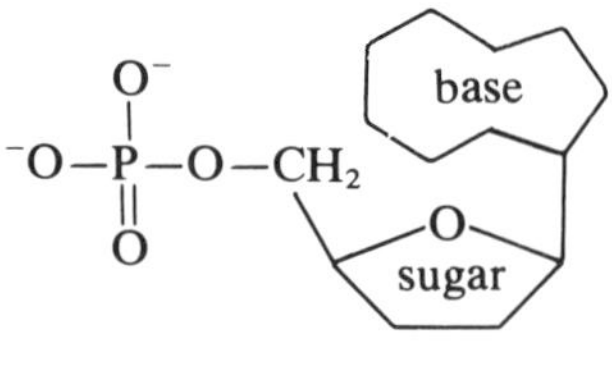

$$^-O-P-O-CH_2$$

a nucleotide unit

Figure 5.5 DNA contains deoxyribose: RNA contains ribose. These are shown above in their straight chain forms. In nucleic acids they occur in the ring forms.

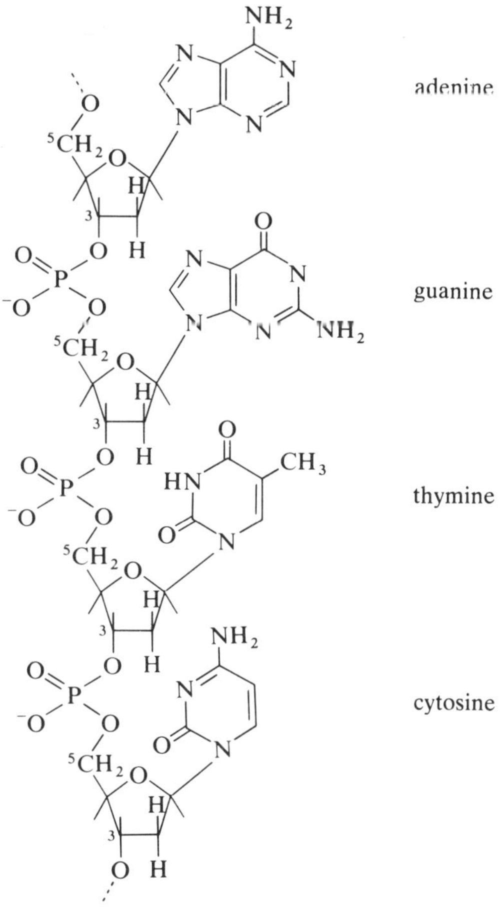

Figure 5.6 Structure of DNA. A short section of a DNA molecule is shown above. Every sugar (deoxyribose) residue has a base attached and the sequence of the bases carries information. DNA molecules may be millions of base—sugar—phosphate units long.

Note that the chain has a 'direction' because of the way in which the sugar residues are linked to the phosphates. Working vertically downwards in the above formula the phosphate is joined to the 5 position of the first deoxyribose which is connected to the next phosphate through its 3 position, and so on. Usually these are referred to as the 3' end and the 5' end. (The prime (') means that we are talking about the numbering of the sugar ring rather than that of the nitrogen-containing base.) Direction is important because when the information is being extracted from the DNA base sequence the enzymes responsible work in one direction only.

Shorthand. It would be tedious to have to write out the complete structure of DNA all the time, especially when all the information is carried in the base sequence. It is usual to write the base sequence simply as a string of initial letters thus:
5'...ATTCGCGTAGGCTTGATCGATGTGC...3'

on. One of the 4 nitrogen-containing bases is linked to every sugar molecule (Figure 5.6). The molecular masses of these giant molecules run into the millions.

Molecules of DNA have a sequence of building blocks. In contrast to proteins where the sequence is of amino acids, the basic building blocks of DNA are base–sugar–phosphate units called nucleotides. Because the sugar–phosphate repeats, the sequence is that of the nitrogen-containing bases. It is important to realize that, like the sequence of amino acids in a protein, the sequence of bases in a DNA molecule is neither random nor repeating.

This much was known in 1950, but almost nothing was known either of the three-dimensional structure of the molecule or of the sequence of bases. To determine the structure of the genetic material had by this time become an irresistible challenge to a number of biochemists.

The double helix

One of the most important discoveries at this crucial time was that it was possible to get an X-ray diffraction pattern from DNA. Maurice Wilkins and Rosalind Franklin in London succeeded in obtaining some reasonably clear diffraction patterns from which it was possible to say that the molecule had regular *repeat* structures: one of 0.34 nm, one of 2.0 nm and one of 3.4 nm. James Watson and Francis Crick in Cambridge then formed what was to be a historic collaboration which resulted finally in the elucidation of the three-dimensional structure of the DNA molecule. Their approach was to take the information that was available at the time, including Wilkins' and Franklins' X-ray data, and use it to try and build a model of the molecule. By keeping bond lengths and angles exact, and by trying to account for the repeats found by Wilkins they hoped that they would at least be able to say that some of the proposed models could be ruled out as being impossible to build, while others were possible.

Watson and Crick believed that the 0.34 nm repeat observed was the distance between successive bases and that the 2.0 nm repeat was the width of the chain. In order to explain the third type of repeat, that of 3.4 nm, they proposed that the chain coiled into a spiral or *helix*. One turn was 3.4 nm from the next and so there would be 10 bases per turn. Looking at such a model, they concluded that it would be much more dense than 'real' DNA, and so they proposed that in practice 2 molecules of DNA 'wound' together to form a *double helix*.

A major problem was the position of the bases in the model. Should they be pointing inwards to the centre of the helix or should they point outwards? Several arrangements were tried on the scale model. The one that best fitted all the known facts was one in which the chains wound in opposite directions (i.e. antiparallel) with the bases pointing *inwards*. Remarkably, if it were arranged that at each point a purine base 'pointed to' a pyrimidine base, a structure rather like a spiral staircase could be built, the base-pairs forming the steps (Figure 5.7). Having a purine pointing towards a pyrimidine allowed hydrogen bonds to form, the effect of which would be to supply the force to hold the chains together and maintain the spiral form.

When models of the 4 nitrogen-containing bases were built and tried opposite each other, only adenine and thymine, and guanine and cytosine, would fit together in the space allowed and form hydrogen bonds (Figure 5.8). Adenine and cytosine would not do this, nor would guanine and thymine.

Figure 5.7 Sketch of a space-filling model of part of a DNA double helix. The overall structure is like that of a spiral staircase with the flat rings of the purine and pyrimidine bases forming the steps (see Fig. 5.8). The repeating sugar—phosphate backbone forms a double spiral with the phosphate groups (charged) pointing outwards. The chains have direction (i.e. $3 \rightarrow 5$) and the double helix is antiparallel, i.e. one chain goes in one direction and the other chain in the opposite direction.

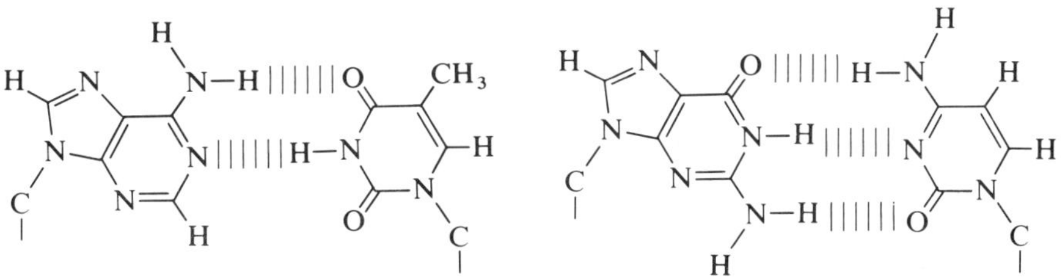

adenine to thymine   guanine to cytosine

Figure 5.8 In the DNA double helix the bases pair by forming hydrogen bonds. Adenine will only match up with thymine (two hydrogen bonds) and guanine will only match up with cytosine (three hydrogen bonds). Note that a purine is always opposite a pyrimidine. These rings are planar or flat so that each of the units shown above can form a step of the spiral staircase (see Fig. 5.7). One strand of DNA will only match up with its *complementary* strand to form a double helix, i.e. the sequence of bases must be such that wherever there is an adenine in one strand there must be a thymine in the other strand, and so on.

This meant that if the sequence of bases along one of the chains of the proposed double helix were specified, then the sequence of its complementary chain was also fixed. A strand of sequence:

$$\text{Pu \ Py \ Pu \ Pu \ Py \ Py \ Py}$$

$$-A-C-A-G-T-C-T- \quad \text{will only 'match' with}$$

$$-T-G-T-C-A-G-A-$$

$$\text{Py \ Pu \ Py \ Py \ Pu \ Pu \ Pu}$$

Here, the bases are simply represented by the initial letters of their names (A = adenine, and so on). By comparing with Figure 5.4 you will see that purine bases (Pu) are always opposite pyrimidine bases (Py). This was very satisfying because it explained the results of analyses Chargaff had performed on DNA from different sources some years previously. He had found that regardless of where a sample of DNA came from and whatever its *total* nucleotide composition, it always contained exactly equal amounts of adenine and thymine, and exactly equal amounts of guanine and cytosine. The reason for this is now clear—this is exactly how the bases must 'pair-up' to form a stable double helix.

Watson and Crick published their ideas on the structure of DNA in 1953, and their model has been consistently supported by later experimental work. It is sometimes said that the elucidation of this structure signalled the beginning of 'Molecular Biology': Watson, Crick and Wilkins were awarded the Nobel Prize in 1962.

Looking at the double helix structure for DNA we can immediately see two important features that fit it for its role as the carrier of genetic information. Firstly, by having 4 different types of base strung out in a long-chain molecule, information can be stored in coded form. The code lies in the *sequence* of the bases in the same way that a string of holes in punched tape represents a meaningful message to a computer (Figure 5.9). Somehow such a code in DNA must carry the information specifying how to join up amino acids in the correct order to make enzyme-proteins. Different types of DNA—different genes—differ from one another *only* in the sequence of bases. We may also note here that just as it is important to feed paper tape into the computer in the correct direction, i.e. not backwards, the code in DNA has a 'direction'. This arises because of the way in which the phosphates are joined to the deoxyribose unit. Enzymes dealing with DNA bind to it in such a way that the code is 'read' in the correct direction.

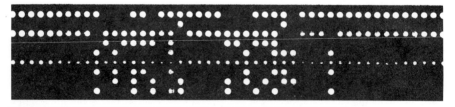

DNA   AATACGTGGATTCGCTAGTTCCGTATAGCTAGATCGGATAGA

Figure 5.9 The information in DNA and RNA is stored as a sequence of bases. This can be 'read' by the appropriate machinery in the same way that punched tape is read by a tape reader of a computer. In both cases the sequence must be 'read' in the correct direction in order to extract the sense of the message

The second important feature of the Watson and Crick double helix model was that it showed how DNA molecules, and hence the genes, could be duplicated at cell division. The next section deals with this important process, called *replication*.

How does DNA duplicate?

At cell division it is necessary to make a replica of the genetic material of a cell so that each of the daughter cells can have a full complement of genes. The way in which the nitrogen-containing bases match-up by hydrogen bonding across the double helix suggested how replication could take place. Suppose that the 2 molecules, or 'complementary strands', of DNA forming a double helix were to be separated and placed in a solution containing a mixture of all 4 nucleotides. Every adenine in a separated strand would hydrogen bond to a free thymine nucleotide, every cytosine to a free guanine nucleotide, and so on, each base to its complement. Now if the bases 'picked up' in this way could be 'zipped up' a new DNA molecule would be created, complementary in structure to the original one. Starting from one double helix we would

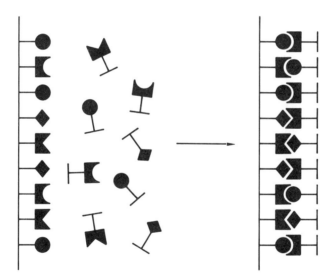

Separated strands of DNA double helix form a template. Free base— sugar—phosphate units are picked up from solution: a purine always matches with a pyrimidine. The bases are held in place by hydrogen bonding—see Fig. 5.8.

New strand of DNA, complementary to template, formed by 'zipping up' by means of enzyme to form new sugar—phosphate backbone. If each strand of the original double-helical DNA does this, two new, identical double helices are formed. Each of these contains one old strand and one newly-formed strand.

Figure 5.10 Replication—forming a new strand of DNA. The mode of replicating DNA is called 'semi-conservative' because each of the new double helices contains one of the original strands of DNA. That this is in fact does happen was shown using isotopes by Meselson and Stahl (Box 5.1)

finish up with two *identical* double helices (Figure 5.10) because each of the original chains would act as a *template*. Thus the original sequences of bases would be duplicated. Although this was originally only a suggestion, a great deal of experimental work confirms that this is, in essence, exactly what does happen. This way of forming a replica of the original DNA double helix is called *semi-conservative replication* because each new double helix contains one molecule of the old DNA double helix and one molecule of the new. A key experiment in support of this mechanism was the Meselson–Stahl experiment (Box 5.1).

Another important discovery, reported in 1957, was that DNA could be made in the test tube using an enzyme called 'DNA polymerase'. This was the enzyme that did the 'zipping up' of nucleotides referred to above. Two interesting features of this DNA polymerase enzyme are important to us here. One is that it only works on the *tri*phosphate forms of the building block nucleotides. These are not ATP, GTP, etc., but a parallel series of compounds which contain *deoxy*ribose instead of ribose and are written 'dATP', 'dGTP', 'dCTP' and 'dTTP'. (Forming a new DNA molecule is a condensation reaction and this is made energetically favourable because these triphosphates are hydrolyzed during the course of the reaction.) The second interesting feature is that DNA polymerase only works if, in addition to dATP, dGTP, dCTP and dTTP, a piece of single-stranded DNA is present to act as a template. When such a template is present in the reaction mixture, the new molecules of DNA formed have not an identical, but a *complementary* sequence of bases, as would be expected.

In practice it seems that when replication takes place in a cell, the DNA double helix does not completely unwind. This would be an extremely difficult operation to perform, both in terms of time and amount of energy required, because of the enormous lengths of DNA found in some cells, e.g. up to 10 cm, normally tightly coiled. It seems that, instead, short stretches of the double helix unwind, new bases come in and are 'zipped up' by DNA polymerase, and then another stretch unwinds, and so on (Figure 5.11). It is possible to 'catch DNA in the process of replicating' in the electron microscope.

Box 5.1 *The Meselson–Stahl experiment*

Watson and Crick *suggested* how replication might take place, and the behaviour of the enzyme DNA polymerase seemed to confirm that a 'primer' piece of DNA was necessary. Proof came from an experiment performed by Matthew Meselson and Frank Stahl in which they very cleverly made use of isotopes. Also, in order to be able to observe several generations of cell divisions, they used bacterial cells which have very short generation times.

They grew their bacteria for several generations in a nutrient medium in which the nitrogen source contained the isotope of nitrogen, ^{15}N. This isotope is not radioactive, but is, of course, 'heavier' than ordinary ^{14}N. The bacteria they eventually obtained contained DNA in which the nitrogen of the bases was labelled with ^{15}N. At this point the cells were suddenly switched to a medium in which the nitrogen source contained ^{14}N. Samples of bacterial cells were taken every generation thereafter and their DNA extracted. Using an extremely sensitive centrifuge method for determining density it was possible to determine whether the DNA samples contained ^{15}N, ^{14}N or both.

Meselson and Stahl found that the original DNA contained, of course, only ^{15}N. After the cells containing this had been through one generation in the ^{14}N-containing medium, however, their DNA was lighter, and appeared to contain half ^{15}N and half ^{14}N. After several generations all the DNA contained ^{14}N only. A possible explanation for these observations is shown in the diagram. Each of the parent strands of DNA stays intact and acts as a template for the formation of a new strand. This is called 'semi-conservative' replication.

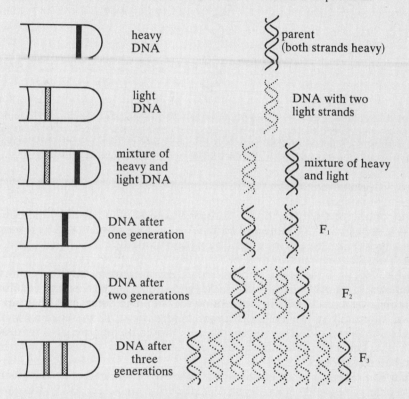

heavy DNA

light DNA

mixture of heavy and light DNA

DNA after one generation

DNA after two generations

DNA after three generations

parent (both strands heavy)

DNA with two light strands

mixture of heavy and light

F_1

F_2

F_3

Meselson and Stahl were able to show that their mechanism was the correct one as follows. They took some of the 'half-and-half' DNA obtained one generation after switching from ^{15}N to ^{14}N medium. According to the theory, each double helix in this should contain one ^{15}N strand of DNA and one ^{14}N strand. By heating the DNA solution and then suddenly cooling they separated the strands. (This procedure disrupts the hydrogen bonds stabilizing the double helix, and the sudden cooling does not allow sufficient time for their reformation with the correct (complementary) alignment of the strands.) Testing the separated strands revealed that half of the strands did indeed contain only ^{15}N and half contained only ^{14}N. No strands were found which contained *both* ^{15}N and ^{14}N, showing that no breaks had occurred in the original strands. Replication was indeed semi-conservative.

How long is a gene?

The Watson and Crick model gives us a very plausible explanation of how information is stored and how replication can occur. It was mentioned

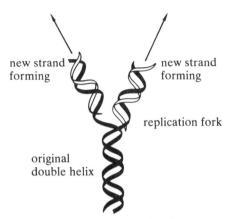

Figure 5.11 During replication it is not necessary for the two strands of a DNA double helix to separate completely. New stretches of complementary DNA can be formed where short sections of double helix become unwound to form a 'replication fork'. This fork can then move along the DNA thus extending the length of new DNA formed

previously that even the simplest cell contains several thousand different types of protein molecule. On the other hand, even the cells of higher animals and plants only contain a few dozen chromosomes. It is fair to ask, therefore, what a gene is: is it a molecule of DNA or a part of one, and how many proteins does it have the information for, one or many?

In 1908, Sir Archibald Garrod, an English physician, gave a series of lectures in which he proposed that certain human diseases were 'inborn errors of metabolism'. He said that certain diseases were caused by a hereditary inability to perform certain chemical processes. When we realize how little was known about biochemistry and of metabolism at that time, it was a remarkable suggestion to make. Because such 'diseases' would be inherited, there would be a direct link between the genetic material and the enzymes catalyzing metabolic reactions.

In the intervening years many such inborn errors of metabolism have been discovered. One well-known example is phenylketonuria, or 'PKU' for short. In this disease, a single enzyme-protein is either missing or, if present, lacks catalytic activity. The function of this enzyme, in the normal individual, is to convert the amino acid phenylalanine to tyrosine (Figure 5.12). This is the normal method of breaking down any excess phenylalanine in the diet, and the tyrosine formed undergoes a number of further catabolic steps. In individuals with PKU the phenylalanine from the diet, in excess of that required for building proteins, accumulates in the blood instead of being broken down. The high concentrations of phenylalanine, and some of its abnormal breakdown products, are harmful to developing brain cells. Children born with this hereditary disease therefore tend to show signs of mental retardation unless the disease is detected soon after birth and the diet changed. The main source of phenylalanine is protein in the diet: in the case of the infant, this is milk protein or casein. Infants with the disease are therefore fed a substitute diet containing all the essentials (amino acids, vitamins, calcium, phosphorus, etc) plus carbohydrate and fat, but containing only just enough phenylalanine for building proteins, so that there is no excess to be

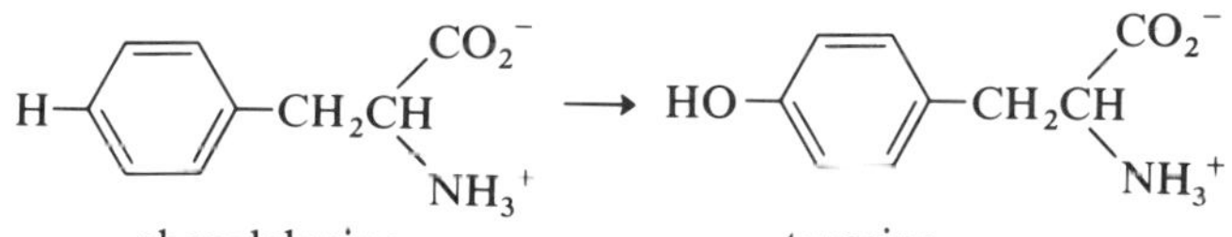

$$H\!-\!\bigcirc\!-\!CH_2\overset{CO_2^-}{\underset{NH_3^+}{CH}} \longrightarrow HO\!-\!\bigcirc\!-\!CH_2\overset{CO_2^-}{\underset{NH_3^+}{CH}}$$

phenylalanine   tyrosine

Figure 5.12 The conversion of phenylalanine to tyrosine involves the addition of a hydroxyl group to the ring of phenylalanine catalyzed by a specific enzyme. This is the first step in the catabolism of any phenylalanine in the diet in excess of what is required for the manufacture of proteins

broken down. PKU is inherited in a Mendelian recessive fashion and about 1 in 15 000 infants is homozygous for the condition, i.e. the trait is passed on from both parents.

The link here is important: an inherited defect appears to result in a deficiency in a single enzyme, although this may result in a number of serious, but secondary, consequences such as mental retardation in this example. Interesting and medically useful as it is to work with human diseases, and human genetics, there are reasons why biochemists have become impatient. Genetic diseases occur only very rarely and long generation times make genetic studies long-term projects. Biochemists therefore turned to micro-organisms in order to find out more about the link between the genetic material and enzyme-proteins. Not only do micro-organisms have short generation times, but it is also possible to produce inborn errors of metabolism, or mutations, which is what these are (see p. 218), in them at will.

The first experiments were performed in 1941 by Beadle and Tatum with the bread mould, *Neurospora crassa* (Box 5.2). It is possible to produce mutations in *Neurospora* by using X-ray irradiation. The organism is haploid for most of its life-cycle, and so any genetic change is expressed immediately because there is no question of it being dominant or recessive.

Beadle and Tatum were able to show that a change in a single gene resulted in a change in a single enzyme. This implies that one gene contains the information specifying the sequence of amino acids in one type of enzyme-protein. A hypothesis was put forward which said, very simply; 'one gene–one enzyme'. This hypothesis is essentially correct and has only had to undergo small modifications over the years since it was first proposed.

Box 5.2 *The experiments of Beadle and Tatum*

The bread mould, *Neurospora*, can grow on a basic nutrient medium containing salts, including nitrates, sulphates and phosphates, and a sugar such as glucose as a source of carbon and energy. Everything else it needs for growth, such as all the amino acids, it makes itself from these simple food supplies. *Neurospora*, therefore, possesses a complement of enzymes capable of making all its amino acids (and many other things too, of course). The question Beadle and Tatum asked was: is the synthesis of all these complex organic molecules under genetic control? The experiments they did were as follows.

Neurospora spores were placed in the simple medium described above and allowed to grow. These spores were exposed to a short burst of radiation with X-rays, following which, single spores were picked out and transferred to a new 'complete' medium containing all the amino acids and all the vitamins, etc. On the complete medium *Neurospora* grows well and does not have to synthesize all its building blocks from scratch. After a period of growth on this complete medium, spores were picked off again. Some were placed on the simple medium and some on the complete medium. All, of course, grew on the complete medium, but some would not grow on the simple medium. Beadle and Tatum concluded that the latter spores had lost the ability to produce one or more of their essential building blocks. If the building blocks were supplied preformed, as in the complete medium, then there was no problem, growth occurred.

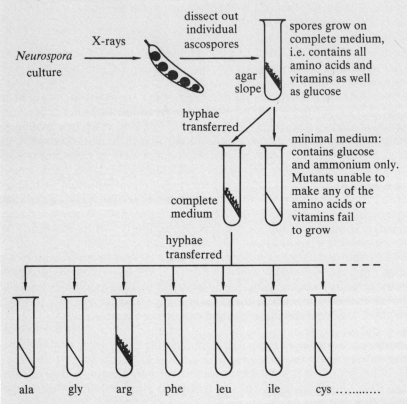

In the last transfer hyphae are placed on agar slopes made with medium containing just one amino acid (plus glucose, plus ammonia). The slope showing growth of hyphae is in this case the one containing arginine. This mutant can therefore make all its amino acid and vitamins *except* arginine. When arginine is supplied, growth takes place.

The experiment was continued by investigating what compounds added to the simple medium, allowed the defective spores to grow. All 20 amino acids and a number of vitamins were tried separately. In most cases it was found that adding just one amino acid out of a possible 20 did the trick and the mould grew. Adding any of the other 19 had no effect. It was therefore concluded that such deficient spores lacked the ability to make for themselves *one* of the amino acids required for making proteins. All sorts of deficiences were found: one unable to make arginine, one unable to make lysine, and so on. These would grow if the simple medium was supplemented with arginine, or with lysine, etc. Beadle and Tatum thus produced a whole collection of mutants, deficient in just one aspect of metabolism. Eventually, they were able to show that the deficiencies were in *single enzymes* making it impossible for the biosynthetic pathway for a particular amino acid to proceed.

These experiments confirmed that one gene was responsible for the synthesis of one enzyme. The X-rays caused mutations by destroying or changing the information carried in the DNA/gene.

'One gene–one polypeptide chain'

Not all the proteins a cell can produce are enzymes. Some hormones such as insulin and glucagon are small proteins, and there are structural proteins too such as collagen and elastin. Are genes responsible for specifying their sequences too? There is no doubt now that the answer to this is 'yes'. Therefore, is it better to say: 'one gene–one protein'—because not all proteins are enzymes? Nearly, but even this is not quite correct.

Many biochemical studies have been carried out with the oxygen-carrying protein, haemoglobin, which again is not an enzyme. Each molecule of haemoglobin is made up of four polypeptide chains, identical in pairs, that is to say two α-chains and two β-chains. A large number of 'inborn errors' have been discovered in which a proper haemoglobin is not formed. Sickle cell disease is one such 'error' that leads to human disease (see p. 214). Such errors, of course, are not 'errors of metabolism' because haemoglobin is not an enzyme. Nevertheless they are mutations and the consequences may be just as serious.

Since a great deal is known about the structure of human haemoglobin, including the complete amino acid sequences of both α- and β-chains, when an inborn error is found, it is possible to ask: what is the error and does it affect *both* chains? It turns out that in many instances the 'error' is to put a 'wrong' amino acid somewhere in the sequence of just one type of chain, either α or β. This may be sufficient to make the haemoglobin behave abnormally and not function properly in carrying oxygen. However, since only one type of chain is affected by the 'error' (and this applies to a number of proteins that are made up of multiple types of polypeptide chains), we can now restate the hypothesis in the form 'one gene–one polypeptide chain'. *We therefore think of a gene as being a section of a DNA molecule whose job it is to specify the order in which amino acids are joined together in order to construct one particular type of polypeptide chain.*

How long are DNA molecules?

What we have said above raises a number of questions such as how many genes are there per DNA molecule, and how many molecules of DNA per chromosome? The DNA of all cells is indeed very long. The DNA of a bacterial cell is typically a single length of double helix with a sequence of perhaps four million base-pairs. The length of this molecule is about 1.4 mm. The largest chromosome of the fruit-fly, *Drosophila melanogaster*, contains a single DNA molecule of length about 2 cm! (Such measurements as these have been quite difficult to make because these enormously long, thin molecules are very easily sheared into smaller segments unless special precautions are taken when handling them.) It is obvious that the DNA molecules that constitute the chromosomes may contain hundreds or even thousands of genes. Packing such long molecules into cells presents problems which are overcome by supertwisting the DNA to form a 'coiled coil'. This in turn presents problems during replication which are solved using special enzymes.

Box 5.3 *Non-chromosomal DNA, circular chromosomes and DNA sequencing*

Until recently the DNA of eukaryotic cells was thought to be confined to the nucleus. However, using both light and electron microscopy along with special staining methods, it has now been shown that there is some DNA in both mitochondria and chloroplasts, usually a single molecule per organelle. Chloroplasts and mitochondria have their own ribosomes and are capable of manufacturing a few proteins. Intact chloroplast 'chromosomes' isolated from the alga *Euglena* are found to be *circular* molecules with a circumference of about 45 μm.

Mitochondria are smaller than chloroplasts and contain less DNA. In mammals the DNA molecule of a mitochondrion is usually circular too, and usually it is supercoiled giving it an average size of about 5 μm. Both these and chloroplast DNA molecules have been termed 'nucleoids'.

In many instances, it is found that the DNA of prokaryotic cells is circular, although this is not invariably true. There had long been the suspicion that the DNA of certain bacteria was circular because geneticists had found that the linkage map for the bacterial genes was circular. That this was the case was confirmed when the molecules were seen in the electron microscope. In the gut bacterium *Escherichia coli*, for example, the DNA consists of two circular, complementary molecules of DNA twisted together to form a double helix. The contour length of this molecule (or really pair of molecules) is about 1.4 mm corresponding to about four million base-pairs. In the bacteria, this molecule is supertwisted—otherwise it would not fit into the cells.

Some virus DNA is circular but some is not. The DNA of bacteriophage T7 is linear, while that from bacteriophage λ is linear when inside the virus particle but circular when inside the host bacterium.

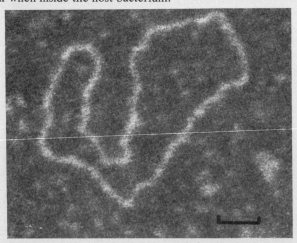

Electron micrograph of circular DNA molecule from a bacterial plasmid (see Figure 6.9). Length of line 0.05 μm. (*Micrograph kindly supplied by Dr P. Kramer, University of Leicester.*)

There has been speculation that the cell organelles of modern eukaryotic cells (e.g. mitochondria, chloroplasts) may be the descendants of bacteria-like organisms that took up residence within primitive eukaryotic cells. Presumably, the presence of these 'endosymbionts' was advantageous to the host cell and their presence was encouraged.

DNA sequencing. The base sequence of the DNA of several bacteriophages is now known and very recently the sequence of human mitochondrial DNA was published (*Nature*, 9th April, 1981). The sequence of bases in human mitochondrial DNA runs to 16 569 base-pairs. It codes for two types of ribosomal RNA, 22 tRNAs, and 13 proteins, and is, of course, circular. This is a remarkable achievement, and only a few years ago would have been thought to be impossible. How was it done?

In the first place it would obviously be a much easier task if the DNA could be broken up into shorter lengths, these sequenced, and then the various bits put together again. This is exactly the strategy used in sequencing proteins (p. 55). However, this could not be done with DNA until a group of enzymes called 'restriction enzymes' was discovered. These are actually 'endonucleases', that is, they split the DNA chain at internal points rather than nibbling away at the ends (Figure 6.5). The availability of restriction enzymes means that large DNA molecules can be cleaved into smaller fragments. These can be separated from one another and then individually treated with a second type of restriction enzyme to yield smaller bits, and so on. In this way, fragments of DNA of a convenient size for sequencing may be obtained.

Finally, methods have been devised for determining the sequences of lengths of DNA a hundred or more bases long. The process involves obtaining single-stranded lengths of DNA, and the building up on these a new, complementary double helix using a polymerase enzyme. While this is happening the radioactive forms of the nucleotides are added so that the newly-synthesized molecule is radioactive. The process is done in the presence of small amounts of an inhibitor, with the result that synthesis stops at random points producing new DNA molecules of a range of lengths. These are then separated from one another by means of gel electrophoresis, and from the position on the gel the base sequence can be read off.

5.4 Protein biosynthesis

So far we have established that the sequence of nitrogen-containing bases along a DNA molecule in a chromosome carries the information, in coded form, specifying the order in which amino acids should be joined to make a protein.

The question we wish to answer now is: how is the coded information in DNA in the nucleus 'translated' into polypeptide chains? It has been known for a long time that although the DNA and the chromosomes are in the nucleus, the biosynthesis of proteins takes place in the cytoplasm.

Where does protein biosynthesis take place?

Almost all of the DNA in a cell is restricted to the nucleus (Box 5.3). Practically all of the protein biosynthesis taking place in a cell goes on in the cytoplasm. Protein biosynthesis does not therefore occur directly using the sequence of bases in DNA as a template. There must be a 'flow of information' from the DNA in the nucleus to the site of protein biosynthesis in the cytoplasm. How does this take place?

It has long been known that tissues highly active in protein biosynthesis contain large amounts of the other nucleic acid, ribonucleic acid, or RNA. Although RNA and DNA have the same overall structure, there are important differences which affect the way in which they function (Table 5.2): RNA tends to exist in the single-stranded form rather than as a double helix. These structural differences are summarized in Figure 5.13.

Table 5.2 Analysis of nucleic acids

DNA is *Deoxyribo*N*ucleic* A*cid*. RNA is *Ribo*N*ucleic* A*cid*.

Content in each case:

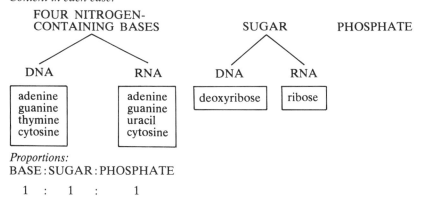

FOUR NITROGEN-
CONTAINING BASES

DNA	RNA
adenine	adenine
guanine	guanine
thymine	uracil
cytosine	cytosine

SUGAR

DNA	RNA
deoxyribose	ribose

PHOSPHATE

Proportions:
BASE : SUGAR : PHOSPHATE
   1 :   1   :   1

In spite of the differences, it is obvious that RNA *could* carry a coded message as a sequence of bases in just the same way that DNA does. When evidence was obtained, using isotopic tracers, that RNA was synthesized in the nucleus, but then moved to the cytoplasmic sites of protein biosynthesis, it became clear that RNA was the *messenger* between the nuclear stores of information and the cytoplasmic protein-building site:

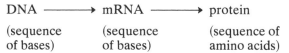

DNA ⟶ mRNA ⟶ protein

(sequence   (sequence   (sequence of
of bases)   of bases)   amino acids)

Enzymes were soon discovered that, given a DNA template and the 4 RNA-type bases (as their triphosphates), catalyzed the synthesis of RNA. This RNA is complementary in sequence to the DNA. Wherever there is a thymine in the DNA sequence an adenine is found in the RNA sequence, and so on. However, wherever there is an adenine in the DNA, a *uracil* occurs in the RNA because of the difference in bases (Figure 5.14). The process of copying, or *transcribing*, the genetic message from DNA to RNA is the first step in protein biosynthesis. Normally only one of the strands of DNA in the double helix is transcribed (the 'transcribing strand'). The RNA transcript can now travel to the cytoplasm for the next part of the process.

In fact *three* types of RNA molecules are synthesized in the nucleus, but only one of these carries the information for making polypeptide chains.

Both nucleic acids have the repeating base—sugar—phosphate structure, but RNA has the sugar ribose instead of deoxyribose:

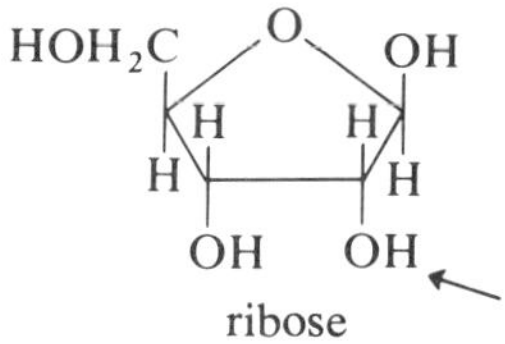

ribose 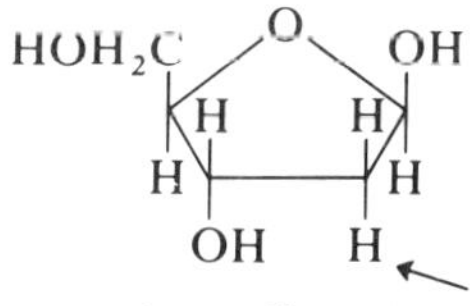   deoxyribose

In addition RNA has the base *uracil* where DNA has thymine:

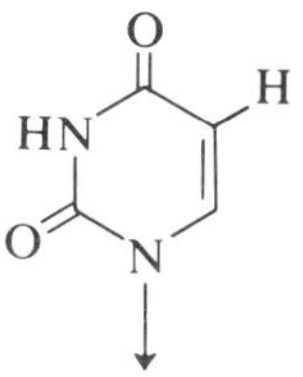

uracil 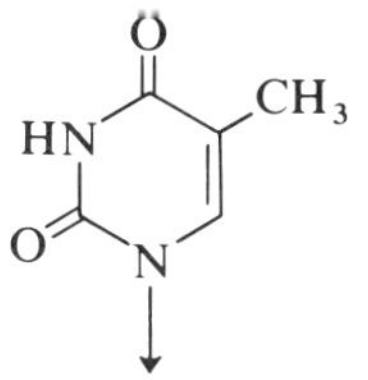   thymine

(The arrows show where the bases are joined to the sugar.)
RNA characteristically tends to be single-stranded rather than forming a double helix between two complementary molecules as in DNA. However, short sections of RNA often complement to form double-helical loops:

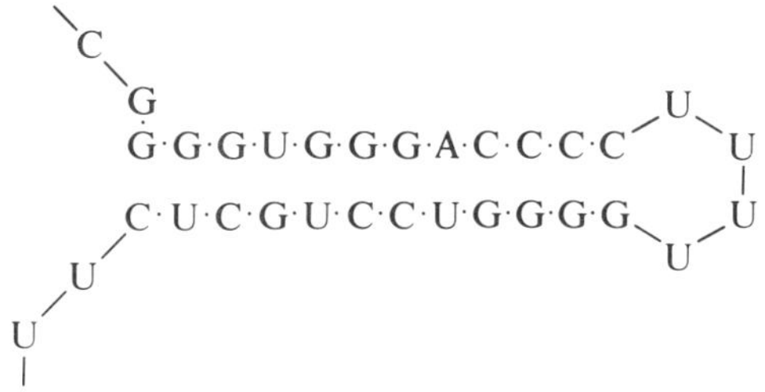

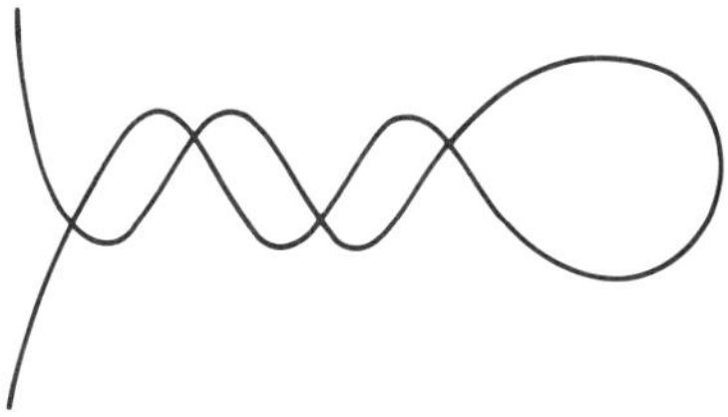

Figure 5.13 Differences between DNA and RNA

Investigation of the other two types of RNA has supplied a great deal of information about how protein biosynthesis takes place. We shall pause here to describe the three types of RNA and their function in protein biosynthesis.

DNA template   RNA strand

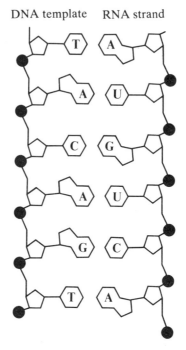

Figure 5.14 RNA strand forming on a DNA template. Note that where the base adenine occurs in the DNA the complementary base in RNA is *uracil*. Note also that the sugar in RNA is ribose rather than deoxyribose, and that the two chains are antiparallel

Messenger RNA (mRNA)

mRNA carries the genetic information as a sequence of bases from the nuclear DNA to the cytoplasm. A given molecule of mRNA may, of course, be long or short depending on whether it carries the coding sequence for a large polypeptide or a small one, but the situation is in practice more complex than this. Sometimes the information coding for a number of polypeptides is transcribed together as a single length of mRNA, with separate polypeptides eventually being produced in the cytoplasm. Often, too, in eukaryotes, additional sequences of bases are linked on to the head and tail ends of the mRNA before it takes part in protein biosynthesis. These additions, 'caps' and 'tails' with long repeating base sequences, carry no genetic information, but have a possible role in initiating protein biosynthesis and/or protecting the mRNA from enzymic degradation in the cytoplasm.

mRNA was the most difficult type of RNA to isolate and study. Not only is it the type of RNA present in the smallest amount in cells (typically about 1% of the total RNA and, of course, existing in a wide range of sizes), but also it is very susceptible to destruction by enzymes and by mechanical damage. One break in the single strand means that the message being carried is lost. We shall return to consider how the 'message' is 'read' and 'translated' later.

Transfer RNA (tRNA)

It has long been known that a small form of RNA is present in the cytoplasm. Previously called 'soluble RNA', this type of RNA, now known as 'transfer RNA', has the function of combining with amino acids to bring them to the site of protein biosynthesis. There is at least one type of tRNA for each of the 20 amino acids found in proteins. Each of the tRNA molecules is composed of about 80 nucleotides in a single strand, and the molecule forms a sort of cloverleaf shape with some internal base-pairing (Figure 5.15). Another unusual characteristic of tRNA molecules is that they contain atypical bases, e.g. pseudouridine. These are formed by means of special modifying enzymes after the tRNA molecule has been assembled from the 4 normal types of nucleotide. A specific region of the DNA in the nucleus carries the sequences that, when transcribed, form the tRNA molecules. These are the 'tRNA genes' but, of course, they do not code for polypeptide chains!

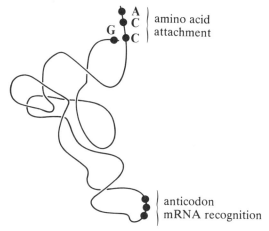

Figure 5.15 Structure of transfer RNA (tRNA). Overall the molecule, about 76 nucleotides long, has a sort of cloverleaf shape. This shape is maintained by short regions of base-pairing within loops. Base number 1 is guanine, and bases 74, 75, 76 are cytosine, cytosine and adenine, respectively. The amino acid attaches to this end of the molecule. The 'anticodon', that is the triplet of bases that complements with the triplet in mRNA, is shown as three black dots at the other end of the molecule, in fact on one of the loops of the cloverleaf

Ribosomal RNA (rRNA)

The third variety of RNA found in cells is ribosomal RNA. Ribosomes are small organelles found in the cytoplasm, and in eukaryotes they are often associated with membranes to form the so-called rough endoplasmic reticulum. Analysis of isolated ribosomes shows them to be composed of proteins and rRNA in very roughly equal amounts. Further investigation of ribosomes shows that each is made up of two subunits. Ribosomes are very complex structures (Figure 5.16) and the function of all the individual components is not fully understood at present. As we shall see, however, the ribosomes are the sites in the cytoplasm where amino acids are joined together to make polypeptide chains. In an actively growing bacterial cell there will be some 15 000 ribosomes present, constituting perhaps a quarter of the mass of the cell. Ribosomes in eukaryotes are similar in form and function to those in prokaryotes, but are somewhat larger.

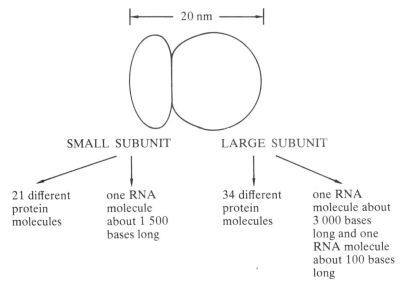

SMALL SUBUNIT		LARGE SUBUNIT	
21 different protein molecules	one RNA molecule about 1 500 bases long	34 different protein molecules	one RNA molecule about 3 000 bases long and one RNA molecule about 100 bases long

Figure 5.16 Ribosomes: a bacterial ribosome consists of two subunits, and each subunit contains both RNA and protein

Bringing together mRNA, tRNA, ribosomes and the 20 amino acids, along with the appropriate enzymes and sources of energy, results in the biosynthesis of polypeptide chains. This indeed can be done in a test tube in a mixture containing neither whole cells nor nuclei. A considerable amount is known about this complex process, but it will be some time before we have a complete understanding of all the steps. Before we look at the process of protein biosynthesis, it is necessary to understand how the information specifying amino acid sequences is carried in the mRNA.

The Genetic Code

Translation is the process by which the coded information in the sequence of bases in mRNA is interpreted to produce a sequence of amino acids in a polypeptide chain. There are 4 types of base and 20 amino acids. Obviously, therefore, one base cannot specify one amino acid. Neither can a set of 2 bases specify one amino acid, because the number of ways it is possible to pick 2 from 4 is 16—not enough to 'code' for 20 amino acids. Assuming that all the codes are the same length, each must therefore be made up of a set of 3 bases at least. The number of ways it is possible to pick 3 from 4, for example, is (4^3) or 64—more than enough to cope with 20 amino acids. It is conceivable that more than 3 bases code for one amino acid. However, this would be inefficient, and evidence suggests that the code is a triplet code (see Figure 5.18).

The realization that the code was a *triplet code* led, in the early 1960s, to intense scientific effort to discover what the code was, that is, to construct a dictionary of 'code words'. By 1964 the code had been elucidated (Box 5.4) and the 'dictionary' is shown as Table 5.3.

Although there are sixty-four possible code words (they are usually referred to as *codons*), and 20 amino acids, it was found that for the majority of amino acids there was more than one codon. Thus the codons ACU, ACC, ACA and ACG all specify the amino acid, threonine. When there are several codons with the same meaning, the code is said to be 'degenerate'. Two amino acids, tryptophan and methionine, are each coded for by one codon. For the rest, which have more than one, it looks as though the first 2 bases are more important than the third. Thus CC (*any base*) always specifies proline. Three of the codons do not specify an amino acid, but are termination signals ('full stops'). Thus at the point at which UAA, UAG or UGA occurs in the mRNA, no further amino acids are added to the growing polypeptide and it is released from the ribosome.

Table 5.3 The Genetic Code

First position	Second position				Third position
	U	C	A	G	
U	phe	ser	tyr	cys	U
	phe	ser	tyr	cys	C
	leu	ser	STOP	STOP	A
	leu	ser	STOP	trp	G
C	leu	pro	his	arg	U
	leu	pro	his	arg	C
	leu	pro	gln	arg	A
	leu	pro	gln	arg	G
A	ile	thr	asn	ser	U
	ile	thr	asn	ser	C
	ile	thr	lys	arg	A
	met*	thr	lys	arg	G
G	val	ala	asp	gly	U
	val	ala	asp	gly	C
	val	ala	glu	gly	A
	val	ala	glu	gly	G

* Also initiation.

Box 5.4 *Deciphering the Genetic Code*

When biochemists started to think about deciphering the Genetic Code in the 1960s, no RNA had been sequenced. It was therefore not possible to translate RNA of known sequence, analyze the polypeptide chain produced and hence find the key to the code. It was, however, possible to make synthetic RNA of repeating sequence. For instance, a piece of RNA having the sequence UUUUUUUUU ... was made. When this was added to a protein-synthesizing system it was found that a polypeptide was manufactured with the sequence phe—phe—phe—phe .... This implied that the codon for phenylalanine was UUU. Similarly, CCC was shown to code for proline and AAA for lysine. Other coding assignments were much more difficult to elucidate, however.

It was eventually possible to decipher the remaining codons using the fact that there is a specific type of tRNA for each of the amino acids. There is, for example, a type for phenylalanine. It was possible to make pieces of RNA only 3 bases long and having a known sequence. Thus, if a piece having the sequence UUU was made, it was found that it would combine with phenylalanine-specific tRNA. Each of the triplet codons was synthesized in this way and tested. From such experiments it was possible to assign the majority of the codons given in Table 5.3.

When the code is written, it is always written in the form in which it occurs in the mRNA. This will correspond to the form in *one* of the strands of DNA (if U is changed to T), and will be complementary to the other strand of the DNA (the strand that has been transcribed).

How is the code read during protein biosynthesis?

The existence of termination codons raises the question of how the code is read and whether there is other punctuation in addition to 'full stops' (termination codons). We also have to decide whether the code is overlapping or not. The two possibilities, overlapping and non-overlapping are shown in Figure 5.17. The weight of evidence is in favour of a *non*-overlapping code. What about punctuation? Are there commas between codons? There is very good evidence that the answer to this question is a straightforward 'No'. Experiments have been performed in which changes in bases are brought about chemically. Such changes are in fact *mutations* (see p. 213), and a very useful one is where one base is actually deleted. Thus a sequence AUGCGUACCUUA ... might become AGCGUACCUUA.... When this happens it is found that complete nonsense results, simply because the 'reading frame' is shifted by one base (Figure 5.18). This implies that apart from 'full stops' mentioned above, there is *no* punctuation. As long as we start reading from the correct base, all is well!

Do all organisms use the same code?

It seems that all organisms do indeed use the same dictionary of code words. This is a strong argument in favour of all modern organisms having arisen from a common ancestor. There are, however, some exceptions that have been discovered recently (Box 5.5).

In cell-free systems used for studying protein biosynthesis it is *only* the mRNA that has specificity. If a suitable mixture of ribosomes, tRNAs and amino acids is prepared, together with an energy supply, the addition of mRNA will lead to the formation of polypeptide chains. For example, if mouse mRNA is used, then mouse-type protein is synthesized. It does not matter that the tRNAs and the ribosomes have come from a rabbit red blood cell or from a wheat embryo. They are non-specific participants in protein biosynthesis which do what the mRNA tells them to do.

The sequence of bases: AUGCGUACCUUAUGGCUG ... could be read in a non-overlapping fashion as follows:

AUG CGU ACC UUA UGG CUG ...

PROTEIN: met—arg—thr—leu—trp—leu ...

Alternatively, if the code were overlapping it could be read as follows:

AUG   met
 UGC   cys
  GCG   ala
   CGU   arg   – which is a completely
    GUA   val   different peptide.
     UAC   tyr
      etc.   etc.

Although an overlapping code would be a more efficient way of storing information (i.e. one base required per amino acid instead of three), it is much less flexible. Once we have started with AUG, the choice of the next amino acid is limited to those beginning UG.., and so on.

The best evidence that the code is non-overlapping comes from the study of mutations. Typically a mutation resulting in a change to a single base in the DNA results in a change of a single amino acid in the corresponding polypeptide. If the code were overlapping such a 'point mutation' ought to affect more than one amino acid.

An exception to this has been discovered in certain very small viruses. In the bacteriophage ϕX174 the gene for 'protein E' is contained entirely within the gene for the larger, completely unrelated 'protein D'. The start signal for protein E is about half way through the sequence for protein D, but the reading frame is shifted by one base. The situation here is one gene–two polypeptides. A point mutation (change of a single base) can result in amino acid changes in *two* polypeptides. The sequence is given below as it is found in the DNA (for the usual codons replace T by U = sequence in mRNA).

ala | glu | gly | val | met | stop   = protein D
... GCGGAAGGAGTGATGTAATGT ...
| arg | lys | glu | stop   = protein E

This arrangement evolved by chance and presumably conferred a competitive advantage on the bacteriophage

Figure 5.17 Is the Genetic Code overlapping or non-overlapping?

Box 5.5 Is the Genetic Code universal?

All the early work on the Genetic Code was done with prokaryotic organisms, but it was soon confirmed that the same code appeared to be used in eukaryotic organisms. Evidence accumulated over the years which very strongly supported the notion that the Genetic Code is universal in all organisms on earth. It was a very satisfying basic concept that all life forms operated on the same language, and, it was argued, that this must be so: any deviation was potentially lethal. In recent years, advances in the technology of sequencing both proteins and genes (i.e. DNA) has meant that it is possible to directly compare gene sequences with gene products. Again there was a general confirmation of the 'universality' of the code.

Very recently, comparison of the sequences of mitochondrial DNA (see Box 5.3) with the corresponding polypeptides, has revealed certain deviations from the apparently universal language, albeit rather small ones. We can say today that the Genetic Code is not universal! The following differences have been found in human mitochondrial DNA (you will find it helpful to refer to Table 5.3 at this point). UGA is read as tryptophan rather than 'stop', AGA and AGG are read as 'stop' rather than arginine, AUA is read as methionine rather than isoleucine, and AUA or AUU is *sometimes* read as a start signal (initiation) rather than AUG. Yeast mitochondrial DNA also appears to show similar divergences.

The code AUG GGU ACC UCU UAU GAU ... is read to produce the polypeptide
met—gly—thr—ser—tyr—asp—. .

Deleting the second base *shifts the reading frame* and produces a completely different protein which will have none of the properties (e.g. enzyme activity) of the original, correct protein:

$$\downarrow$$
AGG GUA CCU CUU AUG

arg—val—pro—leu—met—. .

A deletion elsewhere may result in a termination codon being produced and hence the polypeptide chain may stop at that point:

AUG GGU ACC UCU UAU GAU ... to

AUG GGU ACC UCU UAG AU

met—gly—thr—ser—stop

Deletions have been produced *deliberately* to prove that the Genetic Code is a triplet code. One deletion produces nonsense (as above), two deletions similarly produce nonsense, but three deletions produce an almost normal polypeptide, i.e. it simply lacks a single amino acid, because the reading frame is shifted 3 bases = 1 amino acid: the effects can be seen on a sentence:

NUMBER OF
DELETIONS

0	THE RED HAT FOR THE BAD MAN	sense
1	THE EDH ATF ORT HEB ADM AN?	nonsense
2	THE DHA TFO RTH EBA DMA N??	nonsense
3	THE HAT FOR THE BAD MAN	sense—but a little information lost

Figure 5.18 Effect of deleting one base in a coding sequence

The process of protein biosynthesis

We will now trace the steps by which protein biosynthesis occurs. The tRNA molecules have an absolutely vital role as intermediaries in this process. Because of its structure, a tRNA molecule can recognize its specific amino acid, on the one hand, and a specific codon, i.e. a sequence of 3 bases in mRNA, on the other. The tRNA molecule is the 'interpreter' between the language of the nucleic acids (base sequence) and the language of the proteins (amino acid sequence).

We saw previously that all tRNA molecules are very similar and have a 'cloverleaf' type of structure. One end of the tRNA molecule always terminates with the base guanine (G), while the other end always has the sequence CCA. A series of enzymes exist in cells which are capable of recognizing a specific amino acid and also part of the sequence of the tRNA molecule. (This shows the importance of macromolecules being able to 'recognize' other molecules: recognizing means, as we saw previously, having a complementary shape to the structures to be dealt with.) When, in the presence of one of these enzymes, a specific amino acid and a specific type of tRNA are brought together, the amino acid becomes linked to the CCA end of the tRNA molecule. This process requires the participation of ATP as an energy source.

Another part of the tRNA molecule, in fact part of one of the cloverleaf loops, carries a sequence of 3 bases which form the so-called *anticodon*. Suppose that the tRNA in question was that for the amino acid isoleucine: the anticodon sequence might then be UAU. This is complementary for the codon, AUA, in the mRNA, specifying isoleucine. Recognition of this codon by the tRNA therefore takes the correct amino acid to the site of peptide bond formation which is the ribosome.

Protein biosynthesis, that is the linking of amino acids together, takes place on the ribosomes. The function of the ribosome is to orientate the mRNA and the tRNAs bearing amino acids (charged tRNAs) in a precise way so that peptide bonds form. At present we have too little knowledge of the structure of ribosomes to have any idea of how this takes place. In the electron microscope it is often possible to see groups of ribosomes all reading the same strand of mRNA. These units are called polyribosomes, or usually just 'polysomes'.

The process of protein biosynthesis takes place as follows: following initiation (see Box 5.6) the mRNA is attached to the ribosome. Each three-base unit on this is, of course, a codon specifying a particular amino acid. The tRNA molecule charged with an amino acid, recognizes that this codon 'matches' the anti-codon, and temporarily 'plugs in' to the mRNA by hydrogen bonding. The amino acid is joined to the end of the growing polypeptide chain, and the ribosome moves on 3 bases. The 'empty' tRNA molecule leaves and the next charged one comes in (Figure 5.19). This process continues until a 'full stop' codon is reached. At this point the newly-synthesized polypeptide is dissociated from the ribosome by a special mechanism.

Ribosomes dissociate into their two subunits after the polypeptide chain is released. They appear to reform when a small subunit 'finds' another piece of mRNA to start on, but in fact there is an equilibrium in the cell between associated ribosomes and dissociated subunits. Polysomes of different sizes, i.e. containing different numbers of ribosomes, can be isolated by differential centrifugation: they can also be seen by electron microscopy. Polysomes form because several ribosomes can simultaneously translate one mRNA molecule. In this situation each ribosome is operating independently, each synthesizing a complete polypeptide chain. Polysomes synthesizing haemoglobin polypeptide chains (about 145 amino acids per chain equivalent to about 450 bases in the mRNA) typically have about five ribosomes on one mRNA strand.

In bacteria, the translation of the mRNA starts almost as soon as its synthesis on the DNA has commenced. The end of the mRNA molecule interacts with a ribosome and protein biosynthesis starts while the 'rest' of the mRNA molecule is still being produced. The question of what constitutes the 'start' signal is dealt with in Box 5.6.

(a)

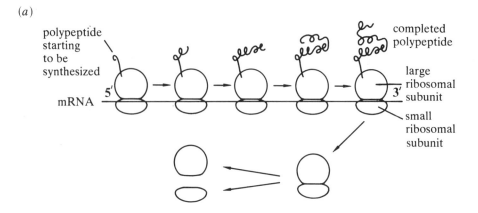

polypeptide starting to be synthesized

5'

mRNA

completed polypeptide

large ribosomal subunit

3'

small ribosomal subunit

(b)

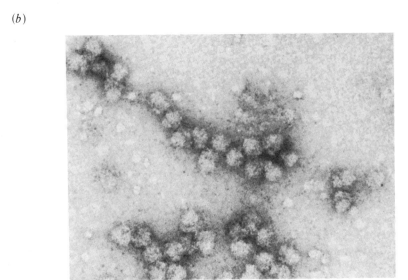

(c)

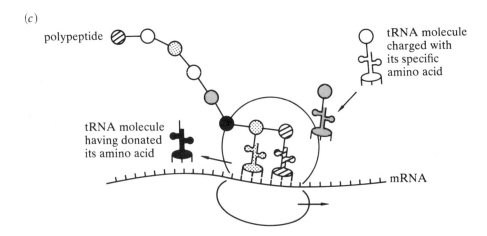

polypeptide

tRNA molecule having donated its amino acid

tRNA molecule charged with its specific amino acid

mRNA

Box 5.6 Initiation of protein biosynthesis

How does protein biosynthesis start? Is the whole of the mRNA molecule translated, i.e. from the very first base so that no start signal is necessary? Evidence suggests that this is not the case. In many types of bacterial mRNA, for instance, up to 5 polypeptides are coded, end-to-end on a single piece of RNA. Each of these has its own start and stop signals.

It has been found that protein biosynthesis in bacteria always starts with the formylation of methionine. This is manufactured enzymically after methionine has become linked to a special tRNA. The tRNA, charged with formylmethionine, somehow associates with a region at the start of the mRNA and with a small ribosomal subunit to form an initiation complex. In the resulting complex the formylmethionine–tRNA complex positions itself so that its anticodon lines up with the initiation AUG codon in the mRNA. This establishes the 'reading frame' for translating the code. A large ribosomal subunit now binds to the complex and protein synthesis can commence.

$$CH_3-S-CH_2-CH_2-CH\begin{array}{c} CO_2^- \\ \\ NH-\overline{CHO} \end{array}$$ N-formylmethionine

Eventually, the formyl group is removed from the polypeptide chain. Sometimes the methionine residue is removed too, but many bacterial proteins do in fact 'commence' with a methionine residue.

When the codon AUG in an internal position in the mRNA specifies methionine, a different tRNA is used, and this is *not* recognized as an initiation point.

A similar although less well-understood system appears to be used in eukaryotic cells. The initiation codon is still AUG but the first amino acid residue (methionine) is not formylated. This methionine is frequently removed during, or following, translation of the protein.

The energetics of peptide bond formation

As mentioned above, peptide bond formation is driven by ATP hydrolysis. The overall reaction can be divided up so that it is possible to understand at which steps 'energy is transferred, and the enzymes that produce tRNA molecules charged with amino acids have a special role here.

First of all in an enzyme-catalyzed reaction, the amino acid reacts with an ATP molecule. Two phosphates are removed at this stage as a pyrophosphate unit, but almost immediately the pyrophosphate is hydrolyzed to give 2

Figure 5.19 Protein biosynthesis. (*a*) Shows the overall process. In the polysome (polyribosome) the ribosomes move along the mRNA in the 5 → 3 direction, functioning independently of one another. When the polypeptide chain is complete it is released and the ribosome splits into large and small subunits. These will reform a ribosome when initiation of protein biosynthesis takes place again at the 5' end of a length of mRNA. Note that proteins are synthesized from the N-terminal end. (*b*) Shows a polysome in the electron microscope: the polypeptides cannot be seen. (*c*) Shows in more detail what occurs on a single ribosome. The cloverleaf tRNA molecules bring amino acids to the ribosome. The three-base anticodon on the tRNA molecule complements with the three-base codon on the mRNA. A peptide bond is formed, the ribosome moves down the mRNA three bases, and a molecule of tRNA (now uncharged, of course) is released. The latter can then go to find another amino acid

molecules of inorganic phosphate. This is effectively hydrolyzing *both* of the pyrophosphate bonds of ATP and the overall reaction is exergonic:

$$\text{amino acid} + \text{ATP} \rightleftharpoons \text{amino acid} - \text{AMP} + \text{P–P}$$

$$\text{P–P} + H_2O \rightleftharpoons 2P_i$$

This 'activation' process is very similar to that occurring in polysaccharide biosynthesis when UDP-glucose is formed, and in fatty acid activation.

The amino acid–AMP complex remains bound to the enzyme until it reacts with the appropriate tRNA molecule, and then the AMP is released.

$$\text{amino acid} - \text{AMP} + \text{tRNA} \rightleftharpoons \text{amino acid} - \text{tRNA} + \text{AMP}$$

The complex formed with tRNA represents a form of the amino acid which is highly labile. Effectively two high-energy bonds have been spent in making it. Eventually when the amino acid is added to the growing polypeptide chain this energy is transferred in part to the peptide bond. In addition, a further expenditure of energy is incurred as the ribosomes move along the mRNA molecule: the hydrolysis of GTP is used to 'drive' this movement.

Processing

The newly-synthesized polypeptide chain may need a little 'finishing off' before it becomes the protein active in the organism. Many 'post-translational modifications' take place and some of them are described in Box 5.7.

Box 5.7 Post-translational modifications

Usually the polypeptide chains formed by the translation of a stretch of mRNA are not the final products. The polypeptide chain may be modified in a number of ways:

(1) The formylmethionine or methionine residues may be removed (see Box 5.6).
(2) Other peptide cleavages may also take place. Certain protein hormones as well as certain virus proteins are synthesized as a very long polypeptide chain which is then cleaved to form several proteins or peptides. Certain enzymes begin life as inactive 'zymogens' from which a small fragment is cleaved to produce an active molecule.
(3) Disulphide bonds may form between two cysteine residues. The specific coiling of the polypeptide chain brings such residues into close proximity.

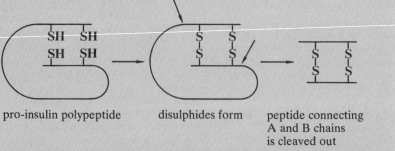

pro-insulin polypeptide   disulphides form   peptide connecting
                                                   A and B chains
                                                   is cleaved out

In the case of the hormone insulin, both cleavage and disulphide formation occur—but in the reverse order. Note that insulin actually has 3 disulphide bonds (not shown here for the sake of simplicity). Full structure shown in Figure 2.9.

(4) In many cases amino acid side chains may be modified to give a protein special properties. In collagen, some proline and lysine residues are hydroxylated. In many proteins, sugar residues are attached to asparagine, serine and threonine residues: this process is called *glycosylation*. Some proteins such as casein of milk are phosphorylated. Prosthetic groups must also be attached to some enzymes.

(5) Subunits must come together if the protein is a multi-subunit protein (e.g. haemoglobin).

Signal sequences. In eukaryotic cells some ribosomes exist free in the cytoplasm while others are bound to the endoplasmic reticulum. Both take part in protein biosynthesis. Secreted proteins such as digestive enzymes are always synthesized by ribosomes attached to the endoplasmic reticulum. There appears to be a special mechanism for getting these newly-synthesized proteins through the membrane so that they are ready for secretion.

It has been proposed that the first few amino acids of these proteins, as they are being synthesized, represent a signal that the whole protein should be secreted through the membrane of the endoplasmic reticulum. As soon as their signal sequence has been made it is recognized by certain proteins found in the membrane. Both newly-synthesized signal and membrane proteins are hydrophobic in character. It seems likely that the signal sequence threads through the membrane as the rest of the protein is being synthesized, the whole polypeptide chain being pulled through the membrane. The signal sequence is then removed by a peptidase enzyme. Several signal sequences have now been determined. Proteins not destined for export do not undergo this sort of processing.

5.5 Mutations

One of the properties of the DNA of a cell is that it must be *stable*. It carries the genetic message from generation to generation and even small changes, as we shall see, may cause the 'wrong' sort of protein, or even no protein at all to be synthesized. This could be fatal to an organism if, for example, it led to the lack of an enzyme essential for metabolism. However, there *is* room for change and indeed evolution depends upon gradual changes taking place in organisms. Most changes will be deleterious, but any that are advantageous may be retained by the evolving organism.

Such changes result from random 'mutations' and we are now in a slightly better position to understand what a mutation is in chemical terms. We can easily foresee that a single base change in DNA will have consequences when protein biosynthesis takes place. If we know the nature of the base change and have a copy of the genetic code dictionary (Table 5.3), we can predict the change in the protein eventually produced.

Many random base changes have occurred in human haemoglobin. Some of the changes have a negligible influence. Thus, changing glycine (GGG) to alanine (GCG) would hardly be detectable because these two amino acids have very similar properties. On the other hand, changes such as that occurring in individuals with sickle cell disease, where a glutamic acid residue is changed to a valine residue, results in the formation of a 'mutant' protein that is much more 'sticky'. The haemoglobin fails to behave normally, and anaemia results (Box 5.8). Equally disastrous consequences would result if an amino acid residue having a key function in the active site of an enzyme were to be changed. Because of the degeneracy of the code, the organism is protected to some extent against the consequences of random base changes. Thus CUU means leucine, but so do CUC and CUA as well as UUG, and so a wide range of change is tolerated in this case. Otherwise, a change from one base to another is likely to change a single amino acid in the resulting protein.

In the cases mentioned above a single amino acid is changed, but a full-length polypeptide is still produced. In other cases, mutations can lead to a failure to produce a polypeptide at all, or to the production of a polypeptide with a *completely different* sequence of amino acids. If a single base change occurred, for example, that changed an amino acid codon to a termination codon, and if this occurred early on in the sequence, then practically no polypeptide would be produced.

It is also possible for a *deletion* in the base sequence to occur. Because there is no 'punctuation' in the code (apart from full stops) this will have the effect of shifting the reading frame (Figure 5.18). This may result in one of a number of consequences. A protein with a completely different sequence might be synthesized. Alternatively, it is possible for a triplet to be read, in the 'new' frame of reference, as a termination codon, with obvious results, as described above. Equally, it is likely that the termination codon at the end of the message will be changed to an amino acid codon. This can lead to the synthesis of a polypeptide longer than the normal one.

The mutations described above all affect the part of the DNA that specifies amino acids. It is also possible for mutations to occur in other parts of DNA such as those that have a controlling function (see below), or in those parts that code for rRNA or tRNA. The changes that result from such mutations will be very much more subtle and difficult to elucidate. They may, nonetheless, still have deleterious or even fatal consequences.

We have concentrated on the deleterious mutations here, but obviously some mutations will be potentially advantageous. Evolution takes place at the molecular level when such mutations make an organism fitter to survive and are retained.

Box 5.8 Sickle cell disease

In 1904, a twenty-year-old black student with a fever was examined by a doctor, James Herrick, in Chicago. The student was feeling weak and dizzy, had a headache and was short of breath. Tests soon revealed that his blood was highly abnormal. Not only was the haemoglobin concentration about half what it should have been, i.e. he was very anaemic, but also, under the microscope it was observed that the red blood cells, instead of being more or less rounded or discoid, were thin elongated structures often with sickle or crescent shapes. Because of this, the disease was called 'sickle cell anaemia', or sickle cell disease.

It was soon found that the disease was far from rare, especially amongst Blacks, of whom as many as four per thousand had the disease. The condition was inherited and usually resulted in an early death. The basic clinical problem was that the oddly-shaped red cells got caught in the capillary blood vessels and easily burst. The average life-time of a red cell was shorter than in normal individuals, hence the development of anaemia. Can we identify the *biochemical* reason for the disease?

We now know exactly what is wrong in sickle cell disease. Individuals with the disease have inherited a mutant DNA in which there has been a change in a single base. As a result of this change an abnormal form of haemoglobin is

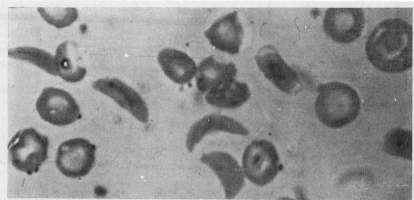

Blood from a patient with sickle cell anaemia. At low oxygen tensions the capsules assume bizarre shapes, become trapped in the small capillaries, and eventually undergo haemolysis.

found in the red cells, called haemoglobin S (normal haemoglobin is haemoglobin A). Haemoglobin S molecules, when deoxygenated, have a strong tendency to stick together to form a fibrous precipitate. Long rods of precipitate deform the red cells, producing the bizarre shapes and making the cells more fragile. Formation of the deoxy form of haemoglobin S is, of course, most likely to happen in the capillaries deep in the tissues.

The biochemical basis of the disease was discovered in 1949 when electrophoresis was performed with haemoglobin S and compared with that of haemoglobin A. Electrophoresis involves applying a spot of haemoglobin solution to a piece of filter paper which has been soaked in buffer solution, and then applying an electric charge of a few hundred volts across the paper. Because the haemoglobin molecules carry a charge they tend to move on the paper towards one or other of the electrodes. A negatively-charged protein molecule will move towards the positive electrode. It was found that, in a buffer solution at pH 8.6, haemoglobin S moved more slowly towards the positive electrode than haemoglobin A.

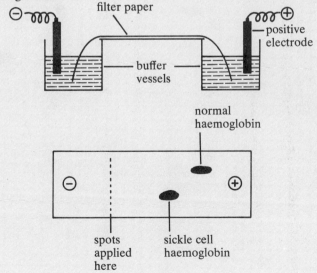

It was concluded that haemoglobin from sickle cell patients had a lower negative charge than normal haemoglobin. The reason for this must be that the amino acids in the two proteins are different. When amino acid analysis was performed it was found that sickle cell haemoglobin had one less glutamic acid residue and one more valine residue than normal haemoglobin. Therefore the change was from glutamic acid (one negative charge at pH 8.6) to valine (zero charge at all pH values). The presence of this valine in haemoglobin S makes the molecule more hydrophobic than normal and this is enough to make the molecules stick together under certain conditions.

Do we know at what position in the polypeptide chain this amino acid substitution has occurred? The answer is 'yes': it is residue number 6 from the amino terminal end of the β-chain. This was demonstrated by the technique of 'peptide mapping' which enabled the complete amino acid sequence of haemoglobin to be determined. Samples of haemoglobin S and haemoglobin A were treated, separately, with the digestive enzyme trypsin. Since this enzyme is specific as to which peptide bonds it will hydrolyze (it cleaves peptide bonds which join the carboxyl group of one of the basic amino acids (lysine, arginine) with the amino group of any other amino acid), only some bonds are split and a collection of small peptides is produced. These will tend to differ from one another in a number of respects (e.g. charge) and may be separated either by electrophoresis or by chromatography. In practice both are normally used, one after the other (see Box 2.4). A spot of the tryptic digest is applied to the corner of a piece of filter paper about 60×60 cm. Electrophoresis is performed in one direction, and then the paper is turned through 90° and chromatography performed, so spreading the peptides over the whole area of the paper. The results obtained, comparing haemoglobin S with haemoglobin A, are shown diagrammatically below:

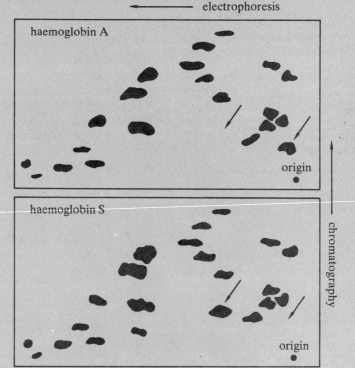

In the peptide map for haemoglobin S one peptide is missing but a new one has appeared (arrowed). It was shown later that this altered peptide contains the first 8 amino acids of the amino terminal end of the β-chain of haemoglobin.

haemoglobin A H_2N—val—his—leu—thr—pro—GLU—glu—lys—...
haemoglobin S H_2N—val—his—leu—thr—pro—VAL—glu—lys—...

Looking back at the Genetic Code (Table 5.3) it may be inferred what the mutation causing sickle cell disease is, namely GAA (code for *glu*) to GUA (code for *val*). A single base change is called a 'point mutation'. Over 300 mutations in human haemoglobins are now known, some causing more serious conditions than sickle cell disease, some causing hardly any distress to the individual.

Medical treatment for individuals with sickle cell disease. What can be done for individuals with sickle cell disease? It is important to realize that some individuals will show worse symptoms than others. Homozygous individuals who have received the mutated gene from *both* parents will be severely affected. Heterozygous individuals who have received one gene for haemoglobin S and one for haemoglobin A are much less affected, to the point of being capable of leading an almost normal life. Examination of their blood shows that both A and S haemoglobins are present. Such individuals may only show signs of distress if they travel in unpressurized aircraft, for example, when the oxygen delivery system of the body is under stress.

For the homozygous individual there is no cure. Treatment consists of repeated blood transfusions to help the anaemia, and this can prolong life considerably. There is the hope in the future that anti-sickling drugs will be discovered. What is needed is a chemical that can penetrate the red cells and interact with haemoglobin S making it less 'sticky', and which also has no harmful effects elsewhere in the body. Some success has recently be achieved with cyanates, $N\equiv C - O^-$. These certainly prevent sickling, but they tend to be too toxic for extensive use in humans. Many other possible compounds are being tested.

How do base changes and deletions occur?

Changes in the DNA base sequence may occur as a result of mistakes during the copying process, or insertions or deletions taking place during crossing over. Such mistakes seem to be comparatively rare events. There are many factors in the environment, both physical and chemical, that can produce base changes. Such factors are called *mutagens* or *mutagenic agents*. Many chemicals are known to be mutagens such as the so-called 'nitrogen-mustards'. X-rays and ultraviolet rays are also mutagenic. Some of these agents have always been present in the environment, but others such as ionizing radiations have increased as a result of man's activities with isotopes and X-rays, as well as his use of chemicals for all sorts of different processes. Be that as it may, mutations are of enormous value to the biochemist in elucidating the way in which life processes occur.

Mutants as tools

In a previous section (p. 94) it was shown how metabolic pathways could be elucidated if metabolic poisons (enzyme inhibitors) could be found that inhibited different steps in metabolism. Metabolic intermediates before the blocked step would accumulate and could be identified. Convenient as this method might seem, it is obviously difficult, if not impossible, to find specific inhibitors for each of the thousands of enzymes involved in metabolism, and so the method has its limitations.

It may have occurred to you by now that instead of using an inhibitor at a particular step in metabolism, one could find a mutant *lacking* the particular enzyme, then the effect would be the same. This gives us a new viewpoint of the inborn errors of metabolism. Many years ago it was found that the urine of certain diseased individuals darkened upon standing, and the disease, shown to be inherited in a simple Mendelian recessive fashion, was called *alcaptonuria*. A similar situation to that found in phenylketonuria applies and, incidentally, is related to tyrosine metabolism. Individuals with alcaptonuria lack one of the enzymes required for the breakdown of tyrosine, and a metabolic intermediate accumulates in just the same way as if the enzyme were blocked by a metabolic poison. It happens that the accumulated product is excreted in the urine and is a substance that darkens by air oxidation. From these clues the pathway of breakdown of the amino acids phenylalanine and tyrosine was worked out. We could say that these human mutations were being used as 'tools' by biochemists in the study of metabolism. (The biochemist would probably agree, but would add that if he can discover the metabolic error, it is likely that he will also be able to propose medical treatment as is the case with phenylketonuria by simply excluding phenylalanine from the diet.) In fact, the use of *microbial* mutants ranks alongside the use of isotopes as one of the most valuable techniques in biochemistry. In addition, there is an enormous industrial potential for using microbial mutants. The first strains of *Penicillium notatum* to be used, for example, produced only small amounts of penicillin. Later, mutant strains were obtained which produced much greater amounts. Many organic chemicals are difficult to manufacture: if mutant microbes can be obtained that produce these and 'excrete' them into the medium, then there is the basis of a useful industrial process—many of the amino acids are produced in this way (see *Biotechnology and Genetic Engineering*, Chapter 6). Because the synthesis is a biological one, typically only one isomer, the L-form of an amino acid, for example, is produced. Chemical methods, in contrast, almost always tend to produce a mixture of isomers which have then to be resolved in further tedious steps.

Mutant microbes may easily be produced by, for example, irradiating with ultraviolet light. Of course, it is not possible to produce specifically one particular type of mutant. A whole range of mutations occur at random: it is then necessary to isolate the required one, but this is comparatively simple to do. With the exception of sickle cell disease, human mutants are, in general, rare, and in any case there may be ethical problems in using them as 'tools'.

5.6 The control of gene expression

Every nucleated somatic cell of a multicellular organism has a full complement of genetic information as a result of the process of mitosis. Cells of an organism take on different forms and functions in order to fulfill their various roles in the whole organism. Thus, although all the cells of an organism have the same DNA, they synthesize a different selection of proteins from all the *possible* ones for which they have the genetic instructions. An extreme example is the red blood cell of vertebrates which synthesizes large amounts of the protein haemoglobin, but very little else. In this section, we ask if it is possible to understand how gene expression is controlled.

This is a very complicated area of biology and we go first to the bacteria where gene expression is reasonably well understood. It must be remembered, however, that eukaryotic cells are immensely more complicated, and have many, many more genes to control. In addition, eukaryotic cells appear to have more DNA than they actually need. The microbial systems give us valuable clues to how gene expression may be controlled, but additional, different mechanisms are likely to operate in eukaryotic cells.

How is gene expression controlled in bacteria?

The DNA of bacterial cells is typically a single molecule or 'piece', and is often circular, corresponding to a single chromosome. Although this codes for several thousand proteins, not all of these are expressed all the time.

Two French biochemists, Jacob and Monod, studied an enzyme system in the gut bacterium *Escherichia coli* (usually called *E. coli*) and the results of their experiments suggested to them a way in which gene expression might be controlled.

E. coli cells are capable of using milk sugar, or lactose, as a nutrient. In order to do this they must split the lactose to its constituent monosaccharides, galactose and glucose, which they can then metabolize. The splitting is done by an enzyme called β-galactosidase (Figure 5.20). *E. coli* cells grown on glucose contain very little β-galactosidase while cells grown on lactose contain considerable amounts of the enzyme. Jacob and Monod found that when they transferred *E. coli* cells from glucose-containing medium to lactose-containing medium, the cells soon started to synthesize β-galactosidase. If the cells were transferred back to the glucose-containing medium, they stopped making this enzyme, which was soon 'diluted out' as cell division continued.

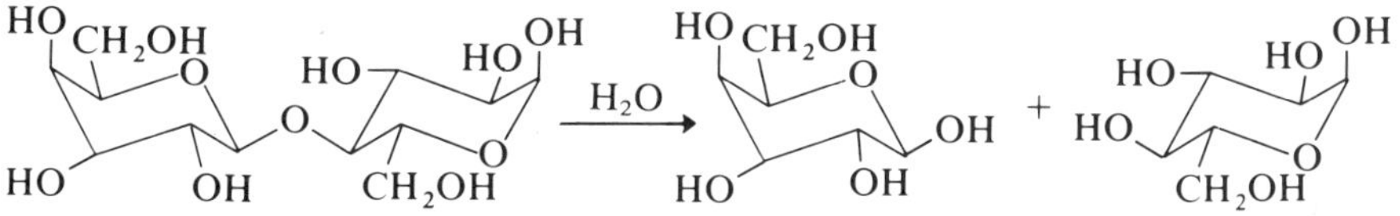

lactose
(= glucose β-galactoside)

galactose

glucose

Figure 5.20 The enzyme β-galactosidase catalyzes the hydrolysis of milk sugar (lactose) to galactose and glucose

From the micro-organism's point of view, this behaviour is very sensible. Why make an enzyme for dealing with a food material if that food material is absent from the medium? In the intensely competitive world of micro-organisms such economic considerations are extremely important. Those cells that can adapt to the prevailing circumstances and save materials and energy in this way are more likely to survive. Even a small saving may be enough to tip the balance in favour of survival. The trick is to have the *information* for making the enzyme, but only to make it when it is needed. In practical terms, it means that on lactose medium each *E. coli* cell contains about 3 000 molecules of β-galactosidase, whereas on glucose medium each cell has, on average, one molecule of β-galactosidase! The question is therefore how does the presence of lactose in the medium 'turn on' the production of an enzyme-protein, β-galactosidase?

Inducers and repressors

Substances such as lactose are called *inducers* because they induce the bacteria to start manufacturing an enzyme to metabolize them. The enzymes whose rate of production they influence are called *inducible enzymes.*

In contrast to this, the 'opposite' process can also occur, that is, the presence of a substance can cause bacterial cells to stop producing a certain enzyme, or even a group of enzymes. Such substances are called *repressors*, and the enzymes they influence, *repressible enzymes.* *E. coli* cells, for instance, are capable of making all the amino acids they need from ammonia and a source of carbon such as glucose. In order to produce these amino acids, perhaps 5–10 enzymes are required, in each case. If one particular amino acid is present in the medium in which the bacterial cells are growing, however, production of the enzymes for its synthesis stops forthwith. The reasons for this are again economic ones: it is inefficient to make enzymes to catalyze processes which are not needed.

You may also note a basic logic here as to whether an enzyme is inducible or repressible. In general, enzymes for dealing with food materials are inducible—they are made only when the particular food material appears in the medium. Enzymes involved in the production of the building blocks for macromolecules such as amino acids are repressible—they are *always* produced unless the building block is available in the medium.

It is found that in bacteria, mRNA molecules are broken down very soon after they are produced. Only a few copies of the polypeptide chain they code for are produced. (This is in contrast to the situation in eukaryotes.) Therefore, control in bacteria, may be exerted at the level of mRNA production. A change in the varieties of mRNA produced from the DNA rapidly leads to a change in the spectrum of polypeptides in the cell. Because the cells divide rapidly a new spectrum of polypeptides can be produced rapidly to meet with the varying compositions of the environments in which the bacteria may find themselves.

The operon

Jacob and Monod proposed a hypothesis to explain the control of gene expression in bacteria. This hypothesis aimed to account for *both* induction and repression within the same framework, and it also needed to account for the fact that control was exerted at the level of the transcription of mRNA from the DNA. Experiments led Jacob and Monod to conclude that *two* genes participated in the expression of β-galactosidase activity. One of these genes, the so-called 'structural gene', carried the sequence of bases coding for the amino acid sequence of the enzyme-protein. The other gene had a controlling influence. It was possible to obtain mutants of *E. coli* in which the mutation was in this so-called 'regulator gene'. The polypeptide chain produced by such mutants was perfectly normal, showing that the sequence of bases in the structural gene was unchanged. However, the enzyme was 'out of control' in these mutants, in that it was produced *whether or not lactose was present in the medium*. Such mutants are called 'constitutive mutants'.

Jacob and Monod proposed that the function of the regulator gene for β-galactosidase was to direct the synthesis of a *repressor protein*. This repressor protein (quite distinct from the polypeptide chain of β-galactosidase) normally inhibits the production of β-galactosidase by combining with a region of the DNA at the point where the production of the β-galactosidase mRNA occurs. If the inducer, lactose, is present however, it combines with the repressor

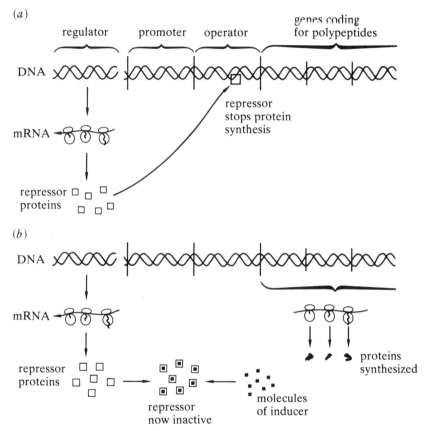

Figure 5.21 Induction of enzyme synthesis. The operon is a group of genes forming a functional unit: often there are several genes coding for the enzyme proteins ('the structural genes') that work in sequence on a particular metabolic pathway. These are all induced simultaneously. In the absence of inducer, a repressor protein is formed using the instructions in the regulator gene and combines with a site on the operator gene preventing protein biosynthesis on the structural genes being initiated. In the presence of inducer, the repressor molecules combine with the inducer and are then incapable of combining with the operator gene. Therefore, biosynthesis of the structural proteins takes place.

The same sequence of events can explain repression of enzyme synthesis, except that now the repressor protein only attaches to the operator gene when it is in combination with the repressor molecule protein causing a conformational change that prevents its binding to this region of the DNA. Hence, β-galactosidase is produced only in the presence of inducer. In the constitutive mutation, an abnormal repressor protein is produced so that control is lost.

Jacob and Monod's hypothesis had to take into account a number of experimental findings and this led them to propose a number of regions on the DNA: *the regulator gene, the promoter, the operator,* and one or more structural genes (Figure 5.21). The stretch of DNA containing the promoter, the operator, and one or more structural genes, is referred to as the *operon*. It was found, for instance, that in both induction and repression, often production of several related enzymes was simultaneously turned on or turned off: normally these were all related to the same metabolic pathway. Thus when an amino acid represses enzyme production, it often represses the synthesis of the several enzymes needed for its manufacture. Repression itself is explained by the fact that in this situation the repressor protein will *only* combine with the transcriptional initiation region on the DNA inhibiting enzyme production if it is in combination with the repressor, e.g. an amino acid in the example mentioned previously. Thus one hypothesis satisfactorily accounts for two 'types' of control, induction and repression, only requiring a slight difference in the properties of the particular repressor protein.

Other control mechanisms in bacteria

Many inducible or repressible coenzyme systems are known in bacterial systems, and, in fact, other forms of control of enzyme synthesis have now been discovered. In addition to these controls, whose function is to prevent the 'uneconomic' production of enzymes not required at a particular time, the types of control mentioned in previous chapters (p. 139) involving allosteric inhibition and phosphorylation are also used by bacteria. Although these forms of control operate more rapidly and exercise 'finer' control, they do not have the economic impact that is achieved by turning off unwanted protein biosynthesis at source, i.e. at the level of the production of mRNA.

What is a gene?

Previously, we have taken a gene to be a region of DNA whose base sequence codes for the sequence of amino acids in a polypeptide chain. This definition has become somewhat eroded by what has been said above, even in the comparatively 'simple' case of prokaryotic cells. When applied to higher organisms it becomes more and more difficult to relate a 'characteristic' such as wrinkled seeds in pea plants, as observed by Mendel, to a particular region of a DNA molecule. As has been said, a mutation in a *control* region can have exceedingly subtle effects which may be difficult to pin down.

Perhaps it would be more accurate to define a gene as a region of DNA that codes for a functional product. This might be a molecule of mRNA that travels to the ribosomes, resulting in the production of a polypeptide chain. Equally it might be a region, as mentioned earlier, that codes for a molecule of tRNA or rRNA. Regions of the DNA whose function it is to combine with a repressor protein, do not code for a functional product, and are referred to as 'regions' rather than genes. In eukaryotic organisms the picture is even more complex and confused, no doubt to a large extent because we just do not understand what large parts of the DNA are for. For example, in the chromosomes perhaps one-fifth of the DNA has no known function and is referred to as 'non-genic', but future research work is likely to cause us to modify our views on this.

How is gene regulation controlled in eukaryotic organisms?

We have a fair understanding of genes and protein biosynthesis and control in prokaryotes. In contrast, we have a very poor idea of what goes on in the eukaryotic cell. We can single out two major problems, which are in fact related to each other. The first is the enormous length of DNA in a typical eukaryotic cell. The longest human chromosome contains about 0.235 pg of DNA, equivalent to a linear double helix of 7.3 cm. At metaphase these chromosomes are observed to be about 0.01 mm long—about 10 000 times shorter than if the DNA were extended. (This applies to prokaryotes too: the DNA of E. coli is equivalent to 1.4 mm length of double helix, but fits into a cell of diameter 10^{-5} mm). How is the DNA packaged, but yet allowed to function? The second problem is that in a highly-differentiated eukaryotic organism, although all of the somatic cells have a complete set of genes, in a given cell type 99.9% of them are turned off most or all of the time. Only red blood cells produce haemoglobin, only pancreatic β-cells produce insulin, and so on. The control of gene expression is extremely effective and efficient, and yet we know almost nothing of how the control is achieved.

The range of DNA contents of cells is very extensive (Table 5.4). Since we know that the information is carried in the sequence of base-pairs, it is perhaps more meaningful to express DNA contents in terms of number of base-pairs rather than in length or as molecular mass. There is an increase in the amount of DNA per cell, more or less in correspondence with evolutionary complexity. However, although the mammals show an unusually compact range (all in the region 3×10^9 base-pairs), some species such as amphibia show an extensive range of sizes from 6×10^8 to 8×10^{11} base-pairs.

Table 5.4 DNA content per haploid cell in different organisms

Species	Mass of DNA in haploid cell (pg)	Number of base-pairs
E. coli (bacterium)	0.0044	4.2×10^6
Yeast	0.02	2.0×10^7
Drosophila	0.18	1.4×10^8
Man	3.4*	3.3×10^9
Onion	16.8	1.6×10^{10}
Lungfish	142.0	1.4×10^{11}

* i.e. about 6 pg in the diploid nucleus of a human cell. Stretched out this would measure about 1.74 m. The cells of one human body contain altogether 16–32 thousand million kilometres of DNA.

DNA and chromosomes

The problem of packaging DNA in eukaryotic cells has special significance, for it is believed that the packing and unpacking of DNA is crucial in making the DNA available to be transcribed to give mRNA. Chromosomes, of course, are only visible during mitosis and meiosis, and in this state the DNA is tightly coiled—and is essentially inactive. In the interphase, the DNA is dispersed within the nucleus and is not easy to visualize by most of the techniques presently available. This is frustrating because this is the time when the DNA is active.

It is believed that each eukaryotic chromosome contains a single length of DNA double helix, i.e. 2 complementary molecules, which is at least a hundred times as long as the *total* DNA of a prokaryotic cell. In chromatin, DNA is always tightly associated with basic proteins called *histones*, and certain less basic 'non-histone' proteins are present too. The DNA, with these proteins, forms highly-ordered arrangements called *nucleosomes* (see below), which have recently been the subject of much research effort. The driving force behind this research has been the belief that an understanding of the structure of the DNA–protein complexes will throw light on the way in which gene expression is controlled in eukaryotes. The suspicion is that although the DNA carries the genetic information, the protein components of the chromosomes controls their structure, and probably their function too.

The combination of DNA with the basic proteins mentioned above is sometimes called 'nucleoprotein', and the association between *basic* proteins and nucleic *acids* largely neutralizes the negative charges on the DNA. Originally, it was thought that the DNA in chromatin was extensively 'covered' in protein, but recent ideas on the existence of a structural unit, the *nucleosome*, has caused a radical change in the last few years in the way chromatin is viewed.

Table 5.5 Five types of histone are found in eukaryotic cells

Type	Relative molecular mass	Lysine : arginine ratio
H1	21 000	20.0
H2A	14 500	1.25
H2B	13 800	2.5
H3	15 300	0.72
H4	11 300	0.79

The histones

In nucleoprotein, a given amount of DNA is associated with approximately the same mass of histone and an equivalent amount of the less basic non-histone proteins. When the DNA replicates, more histone is generated so that the newly-synthesized DNA has the same amount of histone as the parental DNA.

Five types of histone are now recognized as being components of practically all eukaryotic cells (Table 5.5). They are comparatively small proteins (relative molecular mass range 11 000–21 000), but by virtue of their high content of arginine or lysine residues they have a large number of positively-charged side chains (typically one in four residues is either arginine or lysine). Histones from widely-different species are found to be closely similar to one another. In view of the wide variation in DNA sequences, this suggests that histones are concerned with a fundamental aspect of chromatin structure—probably independent of DNA sequence.

The nucleosome

In 1974 it was proposed that chromatin was made up of repeating units, each containing about 200 DNA base-pairs and 2 molecules each of the histones, except H1, that is 8 molecules all told. Such a repeating unit, approximately 11 nm in diameter, is now known as a nucleosome. A chromatin fibre is a chain of nucleosomes, rather like a string of beads, and such fibres can be seen in the electron microscope. Also, if chromatin is treated with the enzyme, deoxyribonuclease or DNase, which splits the DNA sugar–phosphate backbone, particles of approximately 10 nm in diameter are obtained. Analysis of these shows that each contains a length of DNA double helix equivalent to 200 base-pairs.

Further digestion of the nucleosome with a nuclease enzyme trims the DNA down to a 146 base-pair segment of double helix surrounding the core of 8 histone molecules. This is called a 'core particle', and such particles will crystallize. This is fortunate since it means that their structure may be determined by X-ray crystallography! The results of such studies demonstrate that the core particle (Figure 5.22) is a flattish object. The 8 histone molecules form an octamer around which is wound $1\frac{3}{4}$ turns (146 base-pairs) of DNA in a left-handed superhelix. In the nucleosome itself (DNA about 200 base-pairs) there are 'linker' stretches of DNA, joining one nucleosome to the next nucleosome. Since a 200 base-pair length of DNA is 68 nm long, but fits, as a $1\frac{3}{4}$ turn superhelix, into a 11 nm nucleosome, the effective decrease in length, or 'packing ratio' is about 7. This is, by itself, not sufficient to account for the packing ratio in nuclei which is of the order of 1 000! To account for this,

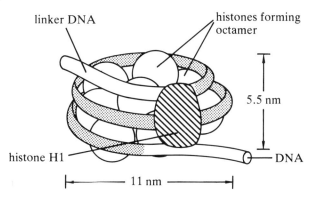

Figure 5.22 A nucleosome. Almost two turns of DNA double helix are wound round a histone octamer. Histone H1 binds to the outside of this 'core particle'. DNA in addition to that coiling round the octamer of histone links one nucleosome assembly to the next. (Redrawn from Stryer: 'Biochemistry', W. H. Freeman & Co.)

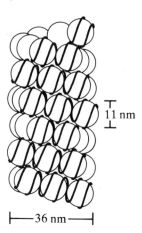

11 nm

├──36 nm──┤

Figure 5.23 Proposed model showing how nucleosomes could form a helical array to achieve a greater packing ratio. There are six nucleosomes per turn of helix (compare with Fig. 5.22). (Redrawn from Stryer: 'Biochemistry', W. H. Freeman & Co.)

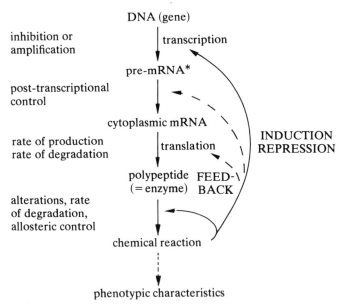

Figure 5.24 There are many points at which control of gene expression may be exerted

* In eukaryotic cells the immediate product of DNA transcription in the nucleus is a precursor of functional mRNA. The precursor is larger than active mRNA and various bits are cut out before the 'edited' molecule reaches the cytoplasm. In addition, a 'cap' is put on the leading end, and a long 'tail' of up to 200 adenines ('poly A tail') is joined to the tail end before the mRNA leaves the nucleus

it has been suggested that there is a further 'order of structure'. It is envisaged that the nucleosomes themselves pack into a helical array (Figure 5.23) some 36 nm in diameter. It has been suggested that histone H1 has the special function of causing the string of beads (nucleosomes) to condense to this helical array. This would give a packing ratio of about 40, and the folding of lengths of helical array into loops would provide additional shortening. Finally, it has been suggested that in each chromosome there is a central protein 'scaffold' to which many long loops of coiled-coil DNA are bound giving the basis for forming the characteristic structure of chromosomes.

Although this proposed structure at present does not tell us how eukaryotic gene expression is controlled, it does provide a scientific framework on which to build hypotheses and design experiments. It represents a considerable advance on previous ideas of chromatin structure, and one gets the feeling that eventually an understanding of the *chemical structure* of the chromosomes will allow us to explain their *biological function*.

In summary, it is likely that gene expression in eukaryotes is controlled at many levels. Figure 5.24 shows the various points at which gene expresion might be controlled.

Questions

1. What is the evidence for:
    (a) the control of cellular activities by DNA,
    (b) replication of DNA,
    (c) nucleotides occurring in matched pairs in DNA molecules?

    [JMB, S]

2. In the late 1950s Meselson and Stahl performed an experiment to investigate the method of replication of DNA and to provide additional evidence relating to its structure as proposed at that time (the double helix of Watson and Crick). They grew bacteria in a medium in which the only nitrogen present was in the form of ammonium chloride containing the heavy isotope ^{15}N. At the end of several generations the bacteria contained ^{15}N in every part of the cell including the nucleic acids. Next, these 'heavy' bacteria were allowed to grow for *one generation* in a medium containing the normal isotope ^{14}N but no ^{15}N. (During one generation each cell divides once.) A sample was removed and its DNA was analyzed. The process was repeated after a second generation. Results are given in the table.

Sample	No. of types of DNA found	Relative weights
'Heavy' parents	one	normal $+ x\%$
First generation	one	normal $+ \dfrac{x}{2}\%$
Second generation	two	(a) normal
		(b) normal $+ \dfrac{x}{2}\%$

    (i) What was the 'double helix' structure postulated by Watson and Crick?
    (ii) What components of the DNA contained nitrogen?

(*iii*) The experimenters were able to explain their results by means of this diagram.

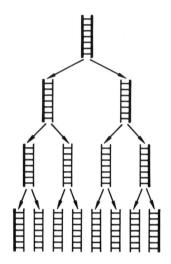

Attempt to relate the diagram to the results given in the table.
(*iv*) Do the results support the Watson–Crick hypothesis?
(*v*) Quote *one* more recent piece of work which has given further knowledge of the details of replication.

[AEB, 1973]

3. The following sequences of amino acids are derived from corresponding peptide chains of the cytochrome *c* of various organisms.

A. man   gly-asp-val-glu-lys-gly-lys-lys-ile-phe-ile-met-lys
B. tuna fish   gly-asp-val-ala-lys-gly-lys-lys-thr-phe-val-glu-lys
C. rhesus monkey   gly-asp-val-glu-lys-gly-lys-lys-ile-phe-ile-met-lys
D. chicken   gly-asp-ile-glu-lys-gly-lys-lys-ile-phe-val-glu-lys
E. horse   gly-asp-val-glu-lys-gly-lys-lys-ile-phe-val-glu-lys

One mutation is assumed for each alteration in a base-pair of the base triplets of the Genetic Code for amino acids. The RNA code is shown opposite.

Example:
The change from Ala(nine) to Lys(ine).

From the table above the code triplets for alanine are GCU, GCC, GCA and GCG, and for lysine are AAA and AAG.
Starting with GCA (Ala), one mutation (G → A) gives ACA and a second mutation (C → A) gives AAA (Lys).
Starting with GCG (Ala), one mutation (G → A) gives ACG and a second mutation (C → A) gives AAG (Lys).
Since in each case two mutations are required the minimum number of mutations is two.

(*a*) Use the table to find the minimum number of mutations which would be required in animals B, C, D and E if their sequences were to be changed to that found in man.
(*b*) Discuss your results.

First letter	Second letter				Third letter
	U	C	A	G	
U	Phe	Ser	Tyr	Cys	U
	Phe	Ser	Tyr	Cys	C
	Leu	Ser	–	–	A
	Leu	Ser	–	Trp	G
C	Leu	Pro	His	Arg	U
	Leu	Pro	His	Arg	C
	Leu	Pro	Gln	Arg	A
	Leu	Pro	Gln	Arg	G
A	Ile	Thr	Asn	Ser	U
	Ile	Thr	Asn	Ser	C
	Ile	Thr	Lys	Arg	A
	Met	Thr	Lys	Arg	G
G	Val	Ala	Asp	Gly	U
	Val	Ala	Asp	Gly	C
	Val	Ala	Glu	Gly	A
	Val	Ala	Glu	Gly	G

[JMB, S]

4. (a) Give a concise description of protein biosynthesis in a cell.
  (b) Describe the structure of proteins under the following headings.
    (i) Primary structure.
    (ii) Secondary structure.
    (iii) Tertiary structure.

[JMB]

5. (a) Explain concisely, those features of the structure of DNA which enable it to:
    (i) serve as a store of genetic information,
    (ii) transmit identical information to the cells produced as a result of mitosis.
  (b) (i) What is a gene and what is a gene mutation?
    (ii) Explain why mutations may appear:
      A. immediately (somatic),
      B. in the first generation,
      C. not until several generations of offspring have appeared.

[JMB]

6. (a) (i) Give a brief and concise account of the structure of DNA (deoxyribonucleic acid).
    (ii) Name *three* types of RNA (ribonucleic acid) and indicate their probable locations.
  (b) Evidence that nucleic acids are responsible for the transmission of inherited characteristics has come from experiments with bacteria, bacteriophages and viruses. Outline any one experiment and say exactly how it provides evidence.

[JMB]

7. DNA is mainly confined to the nucleus of cells and proteins are synthesized in the cytoplasm, yet DNA is said to control protein biosynthesis. Describe the manner in which this control is brought about.
Present and discuss the evidence for it.

[O & C]

8. With the aid of labelled diagrams, describe the structure and reproduction of a named bacteriophage.
Discuss the advantages and disadvantages of this form of parasitism.
In what ways have bacteriophages helped the study of genetics?

[O & C]

9. (a) The wild-type strain of *Neurospora* grows on a minimal culture medium while other strains which have metabolic blocks require the addition of specific compounds to the medium before growth is successful.
   (i) What is meant by the term *minimal medium*?
   (ii) What genetic change gives rise to the different strains?
   (iii) State *one* way in which this same change can be induced in a wild-type strain.

(b) In *Neurospora*, three strains, W, K and T, fail to grow on minimal media.
Strain W will grow if provided with homocysteine or with cystathionine but it will not grow if provided with cysteine.
Strain K will grow if provided with cystathionine, cysteine or homocysteine.
Strain T will grow if provided with homocysteine. It accumulates cystathionine as a result of metabolism.
   (i) From the above evidence, what is the sequence in which these compounds are formed?
   (ii) Mark the position of one metabolic block imposed on any one of the strains, by inserting the symbol ↑ in the above sequence, along with the appropriate letter W, K or T.
   (iii) What is the cause of a metabolic block?
   (iv) Under natural conditions, which one of the four strains K, W, T and wild-type, is most likely to increase in population size?
   (v) What is the term given to the *pressure* acting on the remaining three strains?

(c) (i) In the DNA molecule the only possible base-pair is?
(A.) adenine and cytosine, (B.) guanine and thymine, (C.) adenine and guanine, (D.) thymine and cytosine, (E.) adenine and thymine.
   (ii) The DNA molecule is composed of many nucleotides. Make a labelled diagram to show the arrangement of the three components of one nucleotide.

[Scottish Higher]

10. (a) In which region(s) of a plant would you expect to find a relatively high concentration of deoxyribonucleic acid (DNA) in the cells?
(b) The relative amounts of DNA in certain cells of three species of fish are shown in the following table.

Species	Liver cells	Red blood cells	Sperm
Trout	5.7	5.8	2.8
Carp	3.4	3.5	1.7
Shad	2.0	1.9	0.9

   (i) Explain *briefly* why sperm cells contain about half the DNA of body cells.
   (ii) State *two* other facts which may be deduced from the table.

[Scottish Higher]

11. (a) The following table shows the percentages of each of the bases adenine, guanine, cytosine and thymine present in DNA obtained from different organisms.

Source of DNA	Adenine	Guanine	Cytosine	Thymine
Rat bone marrow	28.9	21.7	20.6	28.8
Herring testes	28.7	20.1	22.1	29.1
Wheat embryo	27.3	22.7	22.8	27.2
Yeast	32.5	18.3	16.8	32.4
Bacterium (*Aerobacter*)	22.0	28.0	28.0	22.0

(i) What evidence is there from the table to suggest that: (A.) base-pairing occurs in DNA, (B.) DNA is universally present in organisms?

(ii) A complete analysis of DNA would show, in addition to the bases, the presence of *two* other components. Name these components.

(iii) What structural change in the DNA molecule may give rise to a mutation, for example, in eye colour, within a species?

(b) The following graph shows the relationship between the numbers of bacteria, the DNA content and temperature in a culture of the bacterium *Salmonella*. The temperature was changed at fixed intervals as shown.

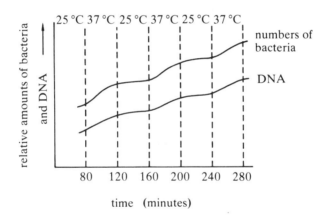

(i) Give *one* reason for the greater increase in numbers of bacteria at 37 °C than at 25 °C.

(ii) Explain the relationship between numbers of bacteria and DNA content of the culture.

(iii) State *one* factor, other than temperature, which could affect the growth of the *Salmonella* culture.

[Scottish Higher]

12. Describe the properties of DNA that allow self-replication to take place, together with one piece of experimental evidence that indicates how this process occurs.
In sickle cell anaemia one of the DNA codons for glutamic acid is changed to that for valine, so the haemoglobin formed does not function in the normal way. From your knowledge of the sequence of events involved in protein synthesis explain why the mutant haemoglobin would be synthesized.

[O & C]

13. Read through the following account of DNA and protein synthesis and then write on the dotted lines the most appropriate word or words to complete the account.
The DNA molecule is composed of .... sugar, phosphoric acid and four types of .... base. Within this molecule the bases are arranged in pairs held together by .... bonds. For example, adenine is paired with .... Adenine and guanine are examples of a group of bases called .... The two strands of nucleotides are twisted around one another to form a double .... , and in each turn of the spiral there are .... base pairs.
The DNA controls protein synthesis by the formation of a template known as .... Compared with DNA, the sugar component of this template is .... , the base .... occurs instead of .... and the molecule consists of a .... chain of nucleotides. The template is stored temporarily in the .... , before passing out into the cytoplasm through the .... It becomes associated with organelles called .... , which supply the .... requred for protein synthesis. Transfer RNA molecules, each with an attached .... , are lined up on the surface of the template according to their .... of .... bases. The amino acids are joined in a chain by .... links to form a polypeptide molecule.

[London]

Further reading

Bauer, W. R., Crick, F. H. C., and White, L. H. (July 1980), 'Supercoiled DNA', *Scientific American*, **243**, 100.
Beadle, G. W. (September 1948), 'The Genes of Men and Moulds', *Scientific American*, **179**, 30.
Cairns, J. (January 1966), 'The Bacterial Chromosome', *Scientific American*, **214**, 36.
Campbell, A. M. (December 1976), 'How Viruses Insert Their DNA into the DNA of the Host Cell', *Scientific American*, **235**, 102.
Crick, F. H. C. (October 1962), 'The Genetic Code', *Scientific American*, **207**, 66.
Fiddes, J. C. (December 1977), 'The Nucleotide Sequence of Viral DNA', *Scientific American*, **237**, 54.
Jackson, R. (1978), 'Protein Biosynthesis', *Oxford/Carolina Biology Reader*, **86**.
Kornberg, A. (October 1968), 'The Synthesis of DNA', *Scientific American*, **219**, 64.
Mirsky, A. E. (June 1968), 'The Discovery of DNA', *Scientific American*, **238**, 78.
Nomura, M. (October 1969), 'Ribosomes', *Scientific American*, **221**, 28.
Travers, A. A. (1978), 'Transcription of DNA', *Oxford/Carolina Biology Reader*, **75**.
Watson, J. D. (1970), *'The Double Helix'*. (Penguin Books).
Watson, J. D., and Crick, F. H. C. (April 1953), 'Molecular Structure of Nucleic Acids', *Nature*, p. 737.
Wood, W. B., and Edgar, R. S. (July 1967), 'Building a Bacterial Virus', *Scientific American*, **217**, 60.
Yanofsky, C. (May 1967), 'Gene Structure and Protein Structure', *Scientific American*, **216**, 80.

6
Biochemistry and Man

Summary

The aim of the final chapter is to demonstrate how biochemical knowledge is relevant to our daily lives and to point to directions in which it will have an increasingly important influence in the future. If one understands a biological process in chemical terms, then it is possible to interfere with that process—to kill a parasite or pest, or to stop a tumour growing. Alternatively, it is possible to use that process in isolation: the prime example here is the employment of isolated enzymes as highly-specific catalysts in the chemical industry. Some of biochemistry's influences on our lives are 'humanitarian' such as those in the area where biochemistry and medicine meet. Thus a knowledge of body chemistry makes it possible to understand the reasons why diseases develop and hence raises the possibility of devising treatments. Other influences are 'economic', though they often have an humanitarian aspect too. Thus the ability to improve food output and to produce energy more cheaply can directly affect the lives of millions in the developing countries. Much of biochemistry's contribution in the future is likely to be within the area called 'Biotechnology'. Included in this is the relatively new technique of 'Genetic Engineering', that is the ability to transplant genes from one organism to another and so control the types of protein manufactured. Very significant developments are confidently expected in this area within the next few years. Armed with a basic knowledge of biochemistry it is possible to understand these developments and indeed to assess the future possibilities.

6.1 Introduction

By comparison with chemistry and physics, biochemistry is a very young science. In spite of its short history, biochemistry has had a considerable influence on human life so far this century and is likely to have an even greater impact in the next decade or two. It is hardly possible to open a newspaper these days without seeing some reference to 'biotechnology' or 'genetic engineering', both of which are based directly on biochemical knowledge. In addition, in view of the effects of man's chemical and biochemical activities on the ecosphere, made daily more acute by an exponentially-increasing population, we ask what the role of the biochemist should be.

233

This final chapter should give the young scientist some idea of the career opportunities that exist in the very broad field called 'the life sciences'. We aim to show that some basic biochemical knowledge is essential for the proper understanding of biology, medicine, pharmacology, genetics, physiology, microbiology and a number of other sciences. The practical consequence of this is that individuals being trained as doctors, dentists, vets, nurses, pharmacists, agricultural scientists, nutritionalists, food technologists and forensic scientists, to name but a few careers, will, during the years at universities, polytechnics and technical colleges, attend courses in biochemistry.

Biochemistry supplies the fundamental chemical knowledge that forms the basis for the life sciences. Armed with an *understanding* of life chemistry, the agricultural chemist, the pharmacologist and the food technologist, are in a position to *interfere*—one hopes intelligently—with a view to improving the quality of life and not polluting the ecosphere. More efficient agricultural practices can be implemented, new drugs with fewer side-effects can be designed, new ways of producing protein to feed a hungry world can be invented, and so on.

6.2 The environment—warfare, pesticides, pollution

In this section we look at the influence of some of man's chemical and biochemical activities on the ecosphere, starting from the perhaps unlikely point of the gruesome effects of modern nerve gases. However, understanding how these work helps us to understand how insecticides work, and even shows us how enzyme catalysis works.

Chemical warfare

Modern chemical weapons, known as 'nerve gases', are based on organophosphorus compounds. 'Tabun', the first of these compounds, was synthesized in Germany in 1936 in the course of research into possible insecticidal compounds, but its military possibilities were soon recognized. The related compound 'Sarin' was discovered in 1938. Sarin is O-isopropyl-methyl phosphono fluoridate:

$$CH_3 - CH(CH_3) - O - P(=O)(F)(CH_3)$$

Although chemical weapons such as hydrogen cyanide and mustard gas (dichlorodiethyl sulphide) were used extensively in World War I, they were not used in Europe in World War II, and indeed the nerve gases have never yet been used in combat. However, the fact that large stocks of these compounds are held by the major powers makes their use in war ever more likely and the possibility of accidents becomes even greater.

Nerve gases, normally stored as liquids, are released as droplets or vapours, and can enter the body by inhalation via the lungs and mucous membranes, or by skin absorption. Their mode of action is to inhibit acetylcholinesterase, an enzyme found in association with the neuronal membrane and which is vital for the normal functioning of the nervous system. When a nerve impulse

labelled peptide amongst a number of unlabelled ones. This peptide was isolated and sequenced and it was shown that the serine occupied position 195 in the polypeptide chain. This serine residue must be at the centre of the active site of the enzyme. Later, it was shown that certain other proteolytic enzymes such as trypsin, elastase (which digests the structural protein elastin) and thrombin (part of the clotting system), *also* had an 'active serine' residue in their catalytic sites.

The next question is 'does acetylcholinesterase have a serine in the centre of its active site?' The answer is, of course, 'yes' and this is why it is so dramatically inhibited by Dip-F and other nerve gases. Hence, there is a family of enzymes, some peptidases, some esterases, whose catalytic activity depends on an active serine residue.

Each of these enzymes is specific: acetylcholinesterase does not hydrolyze peptides, although chymotrypsin and the others show some 'esterase' activity. This indicates that we can think of the active site of any enzyme as consisting of two parts, represented by different amino acid residues. One part is the *catalytic site* and its amino acid residues are actually responsible for promoting catalytic activity or chemical change. The other part is a group of amino acid residues that, by their shapes and charges, etc., attract the appropriate compounds for catalytic action. This is the *specificity site*. Possibly the two sites overlap and one surrounds the other, but the two together form the highly-specific, catalytically-active, 'active site'. Note that not all proteolytic enzymes have an active serine. Pepsin, for example, does not and is unaffected by Dip-F.

Insecticides

Hand in hand with the nerve gases whose use can only be deleterious to mankind, are the pesticides, whose use is aimed at increasing agricultural yields and stamping out insect-carried diseases. There is little doubt that food would be much more expensive if insecticides were not used agriculturally, and that far fewer people now suffer from malaria, etc. because of their use against mosquitoes. However, the success of insecticides must be balanced against the fact that they are not always specific—in addition to killing insects they may also kill other animals, including humans, and upset whole ecosystems. The mode of action of insecticides in biochemical terms is reasonably well understood and the biochemist, along with other scientists, obviously has a role to play in designing safe, but effective insecticides, in testing them sensibly before they are brought into widespread use, and in predicting or treating toxic effects on other animals. Failure to be aware of all the factors involved in a delicately-balanced ecosystem can result in disasters from the use of insecticides of equal or greater severity than that caused by the insect pest it was designed to control.

Factors involved in designing a pesticide

The qualities required of a good pesticide (insecticide or weedkiller) may be summarized as follows:
(1) It should be *specific*; i.e. it should be highly toxic to the pest but non-toxic to other animals and plants, including of course humans.
(2) Having completed its desired action it should degrade rapidly, either in the soil or in the tissues of the pest, to harmless products.

arrives at the synapse, the compound acetylcholine is relea:
the 5–10 nm across the synaptic gap and triggers a new ele(
the post-synaptic membrane so that nervous conduction
acetylcholine is destroyed almost immediately by acetylcholi
acetate and choline. If acetylcholinesterase is inhibited, tl
remains active, triggering nerve impulse after impulse, leading
tion of organs, glands and muscles, an irregular heart rhythm
to convulsions and death.

We can only comment here that the horrendous effects
demonstrates the importance of the balance between ac(
acetylcholinesterase at the nerve junction for the continuatio
happens, a compound closely related to nerve gases, di-isop
fluoridate, or 'Dip-F', has been extremely useful to bioche
revealed how the catalytic sites of many proteolytic enzymes
6.1).

Box 6.1 Dip-F and chymotrypsin

Dip-F was invented as a war gas but in 1949 it was found that it
the digestive enzyme chymotrypsin.

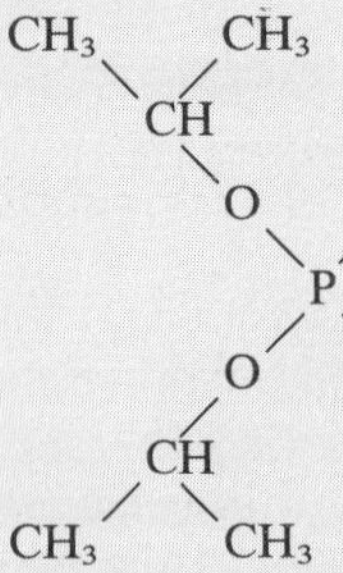

One molecule of Dip-F reacted with one molecule of chymotrypsii
complete loss of catalytic activity. In the process a single serine resid
protein became phosphorylated:

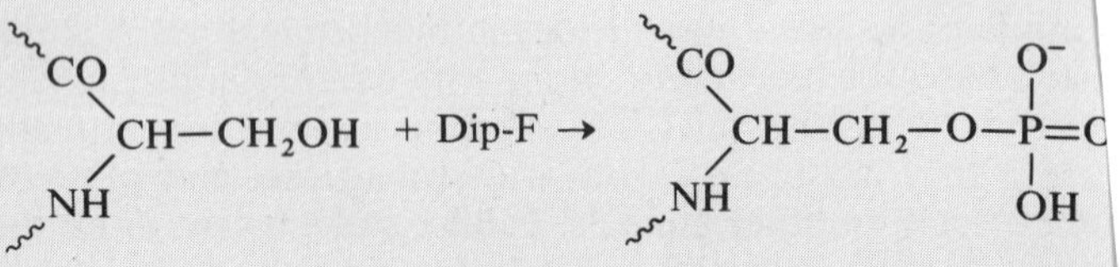

Dip-F has a structure somewhat similar to that of the peptide bonds n
hydrolyzed by chymotrypsin, and is sometimes referred to as a 'pseudosub
The hydroxyl group on the serine residue attacks the phosphorus in
displacing a fluoride ion, and resulting in the formation of a stable phospho
This is the end of the reaction. With a true substrate, i.e. with a pepti
intermediate is formed on the serine which, unlike the phosphoserine, is un:
It rapidly breaks down to give the products of the reaction, leaving the
residue free to participate in another round of catalytic activity.

Not only does Dip-F reveal the mechanism of catalysis, but it also te
which serine residue in the protein is the one involved in catalytic activity. I
Dip-F labelled in its phosphate group with radioactive ^{32}P, it is possib
prepare [^{32}P]-chymotrypsin. Hydrolysis of this with trypsin produces a s

In addition to these scientific and ethical criteria, there are commercial ones too. Not only should the chemical be reasonably cheap to manufacture, but it should be capable of being sprayed, and so properties such as its solubility and dispersibility, and so on, will be important, as will the way in which it penetrates the pest.

Over 200 organic insecticides are presently in use and obviously many thousands of compounds have been tested. Many act by interfering with the electron transport chain. Many other insecticides used today are closely related to the nerve gases described above. They kill insects by inhibiting acetylcholinesterase, but are less toxic to humans and other animals. The reason for this is that insect and mammalian metabolic systems differ in the way in which they 'detoxify' these compounds, and also because certain of these compounds can penetrate insect tissues very rapidly via the tracheal system. Some of these insecticides are not active until they have undergone 'bioconversion', that is, metabolism, in the insect tissues. Here again the metabolic transformation undergone in insect tissues may be quite different from that in mammalian tissues. It is clearly very valuable to know about these metabolic transformation systems when new insecticides are being designed—a very important role for the biochemist in the pesticide industry. An example of 'bioconversion' or 'bioactivation' is found in the case of the insecticides parathion and malathion (Figure 6.1). In this bioconversion a relatively non-toxic P=S group is converted to a highly-toxic P=O group in the tissues of the insect.

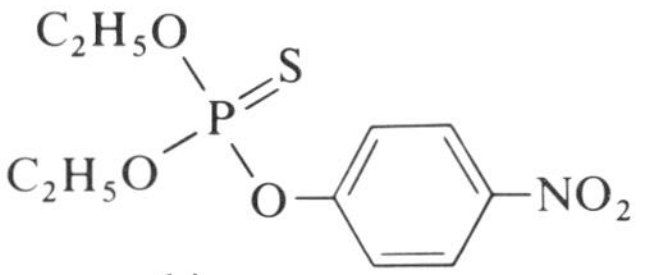

parathion

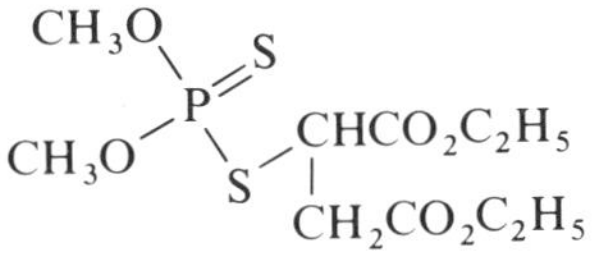

malathion

BIOCONVERSION

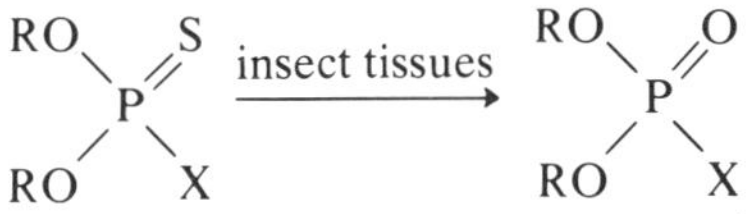

relatively harmless   highly toxic

Figure 6.1 Many insecticides are very similar in structure to the nerve gases that act by inhibiting the enzyme acetylcholinesterase. Parathion and malathion however only become active in the tissues of the insect after 'bioconversion' to a lethal compound

The insecticides that are related to nerve gases are rapidly hydrolyzed in the presence of moisture, and therefore do not persist for very long in the soil, being rapidly converted to harmless compounds. In contrast, the 'chlorinated hydrocarbons'—a group which includes DDT, BHC (benzene hexachloride), dieldrin, endrin, aldrin, chlordane, lindane, and many others—are much less easily degraded and tend to accumulate in animal tissues. DDT is a good example of this group (Box 6.2). Although its precise mode of action is not understood, it is a very effective and cheap insecticide. Unfortunately, it is degraded only to a very limited extent and tends to accumulate in the fatty tissues of animals. This may well lead to its entry into a food chain. In this way, a compound used widely and with every good intention becomes a dangerous pollutant in the environment.

Box 6.2 DDT

*D*ichloro *d*iphenyl *t*richloroethane or 2,2-*bis*(*p*-chlorophenyl)-1,1,1-trichloroethane (DDT) is one of the most effective and inexpensive insecticides, and was first introduced on a large scale during World War II. This compound is lethal to a broad range of insects and has played an enormously important role in increasing food crop production and in reducing insect-bourne diseases such as malaria and typhus. The mechanism of its action is not understood, but is thought to be connected with its solubility in lipids. Possibly, it finds its way into the lipids of membranes and in this way causes hyperactivity, convulsions and eventually death in insects.

So much for the good news. The bad news is that DDT is extremely stable, persists in the ecosphere for years and also accumulates in fatty regions of the tissues of animals including man. There are billions of kilograms of DDT circulating in the biosphere at present even though its usage has been completely abandoned in many areas of the world. It has been estimated that 50% of the DDT sprayed in a single treatment may still be present in a field 10 years later. The other 50% may not have been degraded to harmless products either—but is probably still active somewhere else.

DDT gets into food chains. Predatory birds are particularly at risk because they prey on the birds and small mammals which have consumed DDT-contaminated insects. In these birds DDT accumulates in the liver and fatty tissues: certain species such as the bald eagle and the peregrine falcon are under threat of extinction because of the damage pesticides, such as DDT, are doing to their rate of reproduction. It seems that DDT accumulation in the liver induces the production of enzymes that break down various steroid hormones. This upsets the reproductive system and also appears to influence the calcium transport system. The result of this latter effect is to make egg shells fragile—again with a deleterious influence on the number of offspring produced.

The effect of DDT on humans is not known. Certainly, all populations of vertebrates all round the world contain appreciable amounts of accumulated DDT. Most of us eat animals and human tissues at present contain, on average, about 10 parts per million of DDT and related chemicals. It is already being passed on to the next generation in the fat component of milk.

Herbicides

Many organic compounds have been produced for use as 'weedkillers' or herbicides, and the same criteria apply to these as to pesticides generally. Such compounds have also been used as 'defoliants' by the military. Obviously, their agricultural use must be balanced against their effects on ecological systems, their toxicity and, in some cases, their possible carcinogenic, i.e. cancer-inducing, activity.

Many herbicides act by inhibiting steps in the photosynthetic process. There is great potential here since it should be possible to use compounds which are highly toxic to plants, but completely non-toxic to animals which do not have photosynthetic systems. Also, however, it is necessary for herbicides to be 'selective', that is, to kill the weeds but not the crops. This is obviously more difficult and it is necessary to exploit differences in the metabolic systems of different plants—monocotyledon versus dicotyledon, C_3 versus C_4, and so on. The selective action of many selective weedkillers depends on the fact that some species of plants have built-in detoxification systems whereas others do not. Here again there is a role for the biochemist in finding out about plant metabolism. It must be admitted, however, that, just as happens in the pharmaceutical industry (p. 255), it is often easier and more profitable to screen a whole range of organic compounds for useful ones, rather than to sit down with present day biochemical knowledge and 'design' a compound with herbicidal, pesticidal or other biological activity.

A very successful herbicide is '2,4-D' which is short for 2,4-dichloro-phenoxyacetic acid:

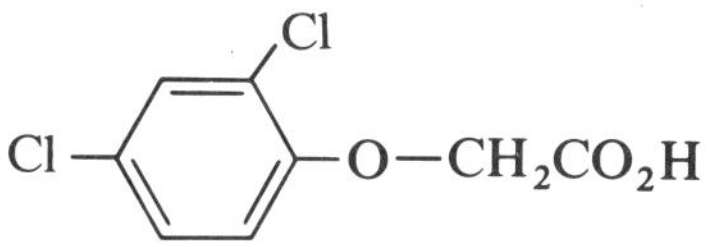

This has an enormous range of effects on the cellular processes of plants. It inhibits certain of the glycolytic enzymes, uncouples oxidative phosphorylation and stimulates protein biosynthesis, for example. Possibly, it does this by mimicking the action of the naturally-occurring plant hormone, indoleacetic acid.

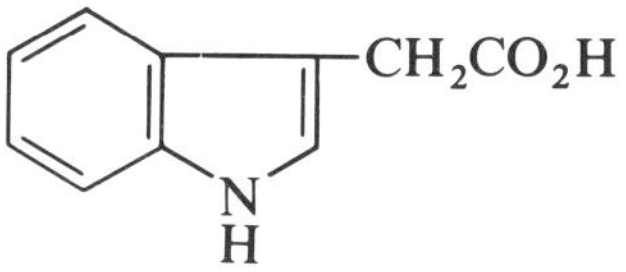

The result is unregulated growth followed by a rapid death.

6.3 Energy and food

The use of insecticides and herbicides has had a major impact on the world's food supply, particularly in terms of improved yields from crops. The use of synthetic fertilizers and of improved strains of plants has also contributed substantially. In spite of this, over the past 50 years or so a major proportion of the world's population has been undernourished or starving. Estimates vary, and definitions of 'undernourishment' vary too, but there is no doubt that about 50% of the world's population is hungry today.

The world population has increased at about 2% per year for the last 10 years and food production has increased at about the same rate. However, in the developed countries (about one-third of the world's population) the annual population is increasing at about 0.7%, and the food production about 2.8%: things are getting better. In developing countries, on the other hand, things are getting worse. The population is increasing at about 2.4%, but food production at only 1.5% per year.

Increases in agricultural production, as mentioned above, depend on a number of factors. In addition to pesticides, fertilizers and plant breeding, there is a need for energy. The production of artificial fertilizers is heavily energy-dependent, and even pumping irrigation water from wells requires energy. Any increases in the world price for oil has disastrous consequences on Third World agriculture.

Energy

All man's energy is eventually derived from the sun. Most of this energy is collected by photosynthetic organisms and trapped as highly-reduced forms of carbon. The average human being requires about 8 500 kJ per day in the form of food to stay alive. This amounts to about $100 \, \text{J s}^{-1}$ or a continuous input of about 100 watts—about the same as an electric light bulb. Considering what a man can do, he is very efficient in terms of his personal energy consumption!

However, there are about 4×10^9 humans on earth, and therefore the continuous energy input required is about 4×10^{11} watts in the form of food. The agricultural production of the world can just about supply this, and could potentially produce a little more, but not much more. We cannot support an exponentially-growing population much longer.

So the overall picture is extremely gloomy for man. What contribution can the biochemist make? The invention of more effective, reliable and acceptable contraceptive drugs could help to reduce demand for energy and food, but there is a social problem too. Apart from any religious strictures, there is the problem of persuading families who are dependent for survival on the agricultural labour of their offspring, to use contraception.

Energy farming

In both the short and the long term there are perhaps more useful contributions the biochemist can make. In the long term it may be possible to copy the chemical mechanisms used by photosynthetic organisms in order to harvest more of the sun's energy. The sun radiates about 1.3×10^{17} watts to the earth—some 20 000 times our present consumption. This energy is available whether we use it or not, and so we are not 'robbing future generations' by using it now: is there not some biological way of trapping more of this energy or 'energy farming'? Some energy is used directly at present, trapped by means of solar panels and the like which trap heat energy, but this is inefficient and costly. At present, large banks of silicon photoelectric cells such as are used in artificial satellites are too costly to be worthwhile. Conceivably it will be possible to devise a chloroplast suspension which, in light, will reduce a synthetic dye, splitting water in the process, and then add a 'hydrogenase' enzyme to convert the reducing equivalents to molecular hydrogen. Hydrogenase enzymes from bacteria are well known. Combustion of hydrogen in oxygen releases the stored energy and does so very cleanly since the product is water. It has been calculated that if such an artificial system, using chlorophyll and hydrogenase, could be devised, operating at 10% efficiency of energy conversion, 20 g hydrogen could be produced per day per square metre of land in south-west USA. Probably, 10% efficiency is wildly over-optimistic.

The mean efficiency over a day for a plant such as maize might be 3%, but for most plants values of less than 1% would be typical. Nevertheless at 20 g hydrogen per day per square metre it would require about an area of 60 km^2 to support a city of a million people.

Energy farming at present cannot be done with artificial systems. It therefore means taking the best advantage of the available energy trapped by photosynthesis. Maize might be grown and harvested for the corn, while the rest of the plant tissue is fermented to produce ethanol to be used instead of petrol. Indeed the increase in oil prices since 1973 has caused many countries to think very seriously about the production of alcohol as a fuel.

'Gasohol'

Up to 20% of ethanol by volume can be added to the petrol used in a car engine without it being necessary to make any adjustment to the carburation system. Alcohol (ethanol) has been produced for human consumption by fermentation of sugars for thousands of years. The sugars used are sucrose, glucose and fructose in the main, but the resulting dilute solution of alcohol contains too much water to burn. There are two problems to be overcome. The first is that it would be advantageous to use starch, cellulose and other cheaper, or waste plant products rather than sugars. Some yeasts can convert starch directly to ethanol, and this process is being improved all the time. Cellulose is more difficult because of the problem of finding species with cellulase enzymes. Such systems will be developed eventually, however. The use of mutants is helpful too. Mutants lacking certain enzymes of the electron transport system, for example, tend to concentrate on anaerobic glycolysis and hence ethanol production, *even* in the presence of oxygen. The microbial cultures generate quite a lot of heat and therefore it is often advantageous to use thermophilic strains or mutants.

The second problem is how to remove water. The best fermentation achievable, using alcohol-tolerant strains of yeast, etc. is about 15% ethanol. Such a solution will not burn. If it is desired to use ethanol as a petrol substitute the percentage has to be increased very substantially. Traditionally this is done by distillation, but this is a highly energy-dependent process and tends to require more energy-input than is eventually recovered when the alcohol is burned! This can be partly overcome—with plants such as maize or sugar-cane—by fermenting starch or sugar and then burning the rest of the plant to supply the heat for distillation.

Food

There are two aspects of the food supply problem. One is supplying enough of the energy-yielding components of food such as carbohydrate and fat. The other aspect concerns the protein content of diets. Because humans, and domestic animals, cannot synthesize some of the amino acids found in proteins, these essential amino acids must be supplied as such or as proteins containing them in the diet. Furthermore, as protein, unlike carbohydrate and fat, is not stored, a daily intake of proteins or amino acids is essential for health. For humans the essential amino acids are: isoleucine, leucine, lysine, methionine, phenylalanine, valine, threonine and tryptophan. In addition, the amino acids arginine and histidine seem to be essential at least to young, growing animals and humans.

The daily intake of protein required to supply the various essential amino acids is about 1 g for every kilogram of body mass. Thus about 75 g, or 3 oz, is enough for the average person—a lot less than is normally eaten by Western man. Not all proteins contain sufficient amounts of all these essential amino acids, however. Plant proteins, especially those from corn and wheat, tend to be somewhat deficient, lacking lysine and being low in tryptophan and methionine. This is very ironical because plant material, and plant protein, is the cheapest and the most efficient to produce. If wheat is grown and the grain fed to cattle for human consumption, the efficiency of energy conversion:

$$\text{sunlight} \rightarrow \text{plant protein} \rightarrow \text{animal protein},$$

is extremely low. Lack of 'first-class' protein containing all the amino acids affects individuals in developing countries very seriously. The disease *kwashiorkor* afflicts millions of children in Africa and Asia. When the child is being fed on mother's milk, all is well as the milk protein contains adequate amounts of the essential amino acids. Soon, however, the child is displaced from the breast by the birth of a new child and has to eat protein from wheat, maize and rice. He then starts to develop the symptoms of protein or amino acid deficiency—kwashiorkor—characterized by retardation of growth, anaemia, failure of pancreatic function, resulting in bloated bellies and diarrhoea, kidney troubles, and, in Blacks, the development of light-reddish coloured hair.

The solution to this problem is to produce 'high-quality' protein *cheaply*, i.e. protein containing adequate amounts of the essential amino acids. As has been said, feeding plant protein to domestic animals is inefficient. The animals spend about 90% of the available energy in keeping warm and walking around and only convert about 10% of it into edible animal protein (meat).

An alternative solution is to use micro-organisms to convert bulk, cheap organic compounds into high-quality protein. This is feasible and is done on a considerable scale industrially at present.

Single cell protein

Many organic compounds are available extremely cheaply and in large quantities. Examples are methanol produced from natural gas, higher paraffins distilled from oil, and the cellulose- and starch-containing wastes from paper mills, sugar factories, and the like. Athough these materials are potential sources of highly-reduced carbon, they cannot be fed directly to animals or humans. However, plenty of micro-organisms can use such materials as carbon sources. It is therefore possible to do the 'biomass conversion' in two stages: (1) feed such materials to micro-organisms, and (2) feed the microbial cells or 'biomass' to animals or to humans. The microbial cells are not just 'protein', of course: carbohydrates, fats, nucleic acids, etc. are present too. Nonetheless dried microbial cells are usually referred to as 'single cell protein', since it is the protein component that is important because it is 'high-quality' protein, containing adequate amounts of the amino acids lysine, tryptophan and methionine to meet the requirements of animals. As a bonus, valuable vitamins are also present too.

In principle this material could be used as human food and this, of course, would be an attractive proposition in a starving world. As it happens the material tends to have a high content of nucleic acids which renders it very unpalatable to humans. Until ways of reducing the nucleic acid content *cheaply* are found, this is not a practical proposition. On the other hand, adding single cell protein to animal feed is perfectly feasible and indeed has major advantages. For example, it reduces the need to feed animals on grains such as barley that *are* acceptable for human consumption and reduces the need to import expensive fishmeal and soybean meal for animal feed supplements. The sole reason for using these imported meals is that they contain good quality protein with adequate amounts of lysine, tryptophan and methionine. 'Single cell protein' may be produced by continuous culture methods on an industrial scale (Box 6.3).

Box 6.3 *Single cell protein*

'Single cell protein' refers to the total cell material of bacteria or yeasts grown in culture on simple, cheap carbon sources such as methanol or oil-derived hydrocarbons. The separated, killed cell material is dried and is sold as an animal feed supplement. The estimated world production of this material in 1980 was about 760 000 tonnes.

The plant operated by ICI Ltd at Billingham uses methanol. ICI already produce vast amounts of this chemical for innumerable uses in industry. It can be made easily and cheaply from natural gas or from coal. A species of bacterium, *Methylophilus methylotrophus*, will grow on a medium containing methanol (as carbon source), ammonia (as nitrogen source), mineral salts and oxygen. Some of the methanol is oxidized to carbon dioxide to provide the energy for converting methanol and ammonia to all the types of biomolecule found in the bacterial cells. This particular organism was selected because it has a high rate of growth, a high cell protein content and produces no toxic products. A diagram of the pressure fermenter used is shown below: The final killed, dried bacterial material is marketed as 'Pruteen'.

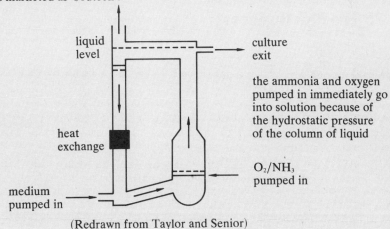

(Redrawn from Taylor and Senior)

A process operated by BP Chemicals Ltd at Grangemouth uses paraffin fractions derived from oil refining and a yeast, *Candida lipolytica*, instead of a bacterium. As would be expected it is necessary to add ammonia, minerals and oxygen, but there are differences compared with the methanol process. Because the paraffins, unlike methanol, are immiscible with water, vigorous mechanical agitation must be used to produce a fine emulsion for the yeast to work on. Yeasts, while growing less rapidly than bacteria, have the advantage that they produce material with a comparatively lower nucleic acid content. Under steady-state conditions it is possible to turn 1 kg hydrocarbon into 1 kg yeast. The material produced is sold as 'Toprina'.

6.4 Biotechnology

'Biotechnology' is a relatively new name given to the use of biological organisms, systems or processes in manufacturing or service industries. In itself the use of biological organisms, especially micro-organisms, industrially or commercially, is not very new. Yeasts have been employed in brewing, fermenting and bread-making for centuries, and for many years now a number of organic chemicals such as citric acid have been made by fermentation processes involving micro-organisms. At the present time, however, the *number* of techniques for using biological organisms and processes in all facets of industry is increasing. The signs are that growth in this area will be even more rapid in the near future in agriculture, the food industry, chemical and pharmaceutical production, metal extraction, waste recycling and effluent treatment, energy production, and so on. Some of the topics already discussed in this chapter, such as the production of single cell protein on a large scale is 'biotechnology' as well as 'big business' (Figure 6.2).

The manipulation of the genetic material, the subject of Section 6.5, is likely to make biotechnology even more important in the next 20–30 years. In this section, we pause briefly to look at some of the ways in which both enzymes and whole organisms are already being used in industry.

9.25 The Risk Business

A Licence to Breed Money

Ask any stockbroker which technologies are the hottest investments today and he'll probably say two words: chips and bugs. For just as the 70s created the chip millionaires, so the 80s are spawning a new breed of bug millionaires. 'Biotechnology' has become the new buzzword for speculators, because biology has come out of the laboratory and into the market-place. Biologists are manipulating genes to produce custom-built bacteria... living factories capable of making some of the most valuable products on earth.

The Risk Business looks at the extraordinary growth of new companies at the frontiers of this technology and sees how they are scrambling to buy up the world's top microbiologists.

Yet there is a South Sea Bubble flavour to this speculation. Promises rather than products have raised millions. The science still has to prove its potential. And there is a more fundamental worry; some scientists fear that big business could alter the course of biological research for ever.

Reporter JUDITH HANN

Film editor ANN LALIC
Executive producer JONATHAN CRANE
Producers OLIVER MORSE, DAVID DUGAN

Figure 6.2 Biotechnology is likely to have a profound influence on our lives in the next decade or two. Apart from the scientific and technological aspects, there are medical, ecological and financial aspects. The above appeared in *Radio Times* of 9th May 1981 advertising a programme on BBC television. (*Reproduced by kind permission of the Editor, Radio Times*)

Enzyme engineering

The chemical industry is highly dependent upon the use of catalysts of various types to speed up processes or to make processes feasible at economic temperatures and pressures. The variety of compounds used as catalysts is very wide, but it always has to be accepted that there will be numerous by-products as well as the desired products of a given reaction. It is not surprising therefore that chemists have looked to biological catalysts, enzymes, to provide highly-specific catalysis in industrial processes. There are many advantages: there will be few by-products and yields will be high, the reaction will proceed at low temperatures and pressures, and high rates of reaction may be expected. The lack of any appreciable amount of by-products is extremely important. In many transformations of organic chemicals it is possible and highly likely that the product will be a mixture of stereoisomers. For many uses a single isomer is required, and the separation of closely similar isomeric compounds can be tedious, time-consuming and expensive. Enzyme catalysis, on the other hand, usually results in the production of a single isomeric form which will typically be the 'biologically-active' form. This is especially important in the pharmaceutical industry where compounds with biological activity as drugs are being synthesized—another reason for enzymes being attractive as industrial catalysts. Before setting up an industrial process involving the use of enzymes there are two questions to be asked:

(1) Is there an enzyme that will catalyze the desired reaction?
(2) How can the enzyme protein be got rid of at the end of the process?

The first problem depends not only on whether an enzyme exists in nature, but also whether an inexpensive source of it can be found. This problem usually resolves itself into the question of whether the required enzyme is produced by easily-grown micro-organisms. If so, then it is usually possible to grow up high-volume cultures of the organism and extract the enzyme. It is especially helpful if the micro-organisms secrete the enzyme into the growth medium. This makes the task of purifying the enzyme protein very simple and the product that much cheaper. An example of the use of enzymes on an enormous scale in industry is the production of fructose syrups for use as sweetening agents (Box 6.4).

Box 6.4 Fructose syrup production

Starch is available in vast quantities extremely cheaply. The molecule of starch consists entirely of glucose residues and if these could be released easily and cheaply the free glucose would find many uses. For example, glucose, as a sweetener in foods, could potentially be much cheaper than cane or beet sugar, sucrose. If, in addition, the glucose could be wholly or partly converted to fructose, the sweetening properties would be greatly enhanced, as fructose is, gram for gram, sweeter than either glucose or sucrose.

Acid has been used in the past to hydrolyze starch to glucose, but more often today bacterial or fungal enzymes are used. In addition, it is possible to convert glucose *enzymically* to fructose: to do this chemically would not be feasible. The material finally produced as 'fructose syrup' is a mixture of glucose and fructose. In fact, different uses in the food industry demand different glucose: fructose ratios. A typical industrial process is outlined in the figure: three steps and four enzymes are employed.

(1) *Liquefaction.* A starch slurry in water (about 30% by weight of starch) is heated to over 100 °C and treated with a heat-stable amylase from the bacterium *Bacillus licheniformis.* This enzyme hydrolyzes a few of the glycosidic bonds in the starch, reducing the viscosity of the slurry, and producing dextrins. Although the enzyme is thermostable, it is eventually destroyed by the high temperatures used within a few minutes. Nevertheless, it has done its job in that time and the concentrated dextrin solution is much easier to pump than the starch slurry.

(2) *Saccharification.* This means 'production of sugar'. The dextrins are further broken down and this time two enzymes are used in order to achieve an efficient and rapid conversion to glucose. A fungal α-amylase is added which converts the dextrins largely to maltose (glucosyl glucose), and an amyloglucosidase that converts the maltose to free glucose. This part of the process is operated at 50–60 °C. These temperatures may seem high compared with physiological ones, but the industrialist is interested in getting the reaction going as quickly as possible. Extensive research has been done to find the highest temperature that may be used (and consequently the highest rate of reaction) coupled with an acceptable rate of enzyme denaturation by heat.

(3) *Isomerization.* In the last step, a part of the glucose is isomerized to the relatively sweeter-tasting fructose using the enzyme glucose isomerase. This again may be operated at near 60 °C and conversions of the order of 30–50% are achieved. In more recent processes this conversion has been achieved by the use of immobilized enzymes (see text).

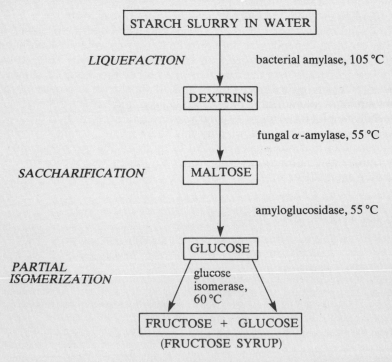

In industrial processes of this type, thousands of tonnes of syrup are produced annually. Enzymes are used by the kilogram, if not by the tonne. Even a small percentage improvement in yield or degree of conversion, by virtue of a small change in process conditions, may prove worthwhile commercially.

When the enzyme has done its job it must be removed from the reaction mixture. How can this be achieved, bearing in mind that every additional step in the process costs money and that the enzymes themselves are expensive? The obvious way of removing the enzyme protein is by heating or acidification in order to denature and precipitate the protein. It may then be removed by filtration. This is done in many processes, but it has been realized that if it could be avoided, money might be saved. Economic pressures such as these caused biochemists to devise new ways of using enzymes in industrial processes. Was it possible, it was asked, to *immobilize* the enzyme-protein so that the substrate could be passed through a 'column' of enzyme particles, being converted to product as it flowed through? Indeed this was found to be feasible. We know that enzyme-proteins are large molecules having on their surfaces a number of functional groups from the amino acid side chains ($-CO_2^-$ and $-NH_3^+$, for example.) The active site only represents a small area of the surface of the protein and the rest of the surface is not involved in catalytic action. A chemical reagent may therefore be used that couples, say, an amino group at a region *remote* from the active site, to an insoluble resin bead (Figure 6.3). The 'enzyme beads' can then be packed into a column and the substrate solution flowed through.

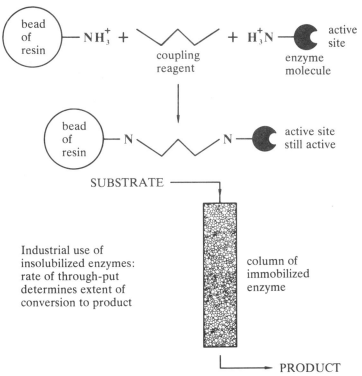

Figure 6.3 Making and using insolubilized enzymes

Such a process is used in the production of fructose syrups from starch as described in Box 6.4. The glucose isomerase used in the last step is immobilized on resin beads which are packed into reaction vessels. The enzyme in this form has a half-life of about 800 hours, and therefore a reaction vessel needs to be changed only after about a month of continuous operation. Furthermore by changing the time for the material to flow through the reaction vessel the degree of conversion of glucose to fructose can be controlled. A plant producing fructose syrups may have a through-put of 1 000 tonnes a day.

There are many more such industrial processes, and it is obvious that from the industrialist's point of view it is worthwhile spending money on biochemical research both *basic*—to find new enzymes and sources of enzymes, and to study their properties—and *applied*—to discover the best ways to use the enzymes, insolubilize them, and at what rates, temperatures and pH values to operate the processes.

6.5 Genetic engineering

Biotechnology is likely to have a major impact on our lives in the next 20–30 years, and one of the areas in which it promises to exert great effect is in the manipulation of the genetic material. Not only can biological organisms and processes be used industrially by man, but also these can be controlled and modified by changing the genetic make up of cells. In this section we discuss how this situation has arisen literally since 1953 when the structure of DNA was discovered.

In a way, man has for many years manipulated the genetic material of biological organisms by plant breeding and animal husbandry. By selection and appropriate crossbreeding, desirable characteristics have been brought out and undesirable ones suppressed. More recently the use of microbial mutants, produced randomly but then selected, has become a routine procedure not only for use in biochemical research, but also for use in industrial processes as described above. Many amino acids and other compounds of biological and medical importance are now produced by micro-organisms grown on a large scale. Typically, mutants are selected which are 'de-repressed', i.e. they have lost control of the expression of certain genes (p. 220). Thus the normal or 'wild-type' organism might only make a given amino acid if it is *absent* from the growth medium. The de-repressed mutant continues to make this amino acid *regardless* of whether it is present in the medium and actually produces so much that it 'excretes' it into the growth medium.

The realization that the Genetic Code is universal (or nearly so, see Box 5.5) has an important consequence—it means that if a genetic message, i.e. a set of genes, can be 'transplanted' from one cell type to another by some means, then the protein-synthesizing machinery of this cell should be capable of translating the message. If, for example, the gene for human insulin could be planted in a bacterial cell, the bacterium could potentially synthesize human insulin. This has, in fact, been achieved recently in the laboratory and there are proposals for it to be done on an industrial scale. What biochemical knowledge has made this possible and what are the prospects for the future? The use of microbial systems for making human or animal proteins is obviously very exciting from the industrial and medical point of view.

Recombinant DNA technology

Organisms have been 'exchanging DNA' for years. Griffith's experiments

(Figure 5.1) on bacterial transformations showed that DNA can pass from one bacterial cell to another, and when fertilization takes place during sexual reproduction in eukaryotes there is a 'mixing of DNA' from two different cells. Now these transfers take place *within* species or between closely-related species, and one of the important biological barriers is that between species. Nevertheless, the Genetic Code *is* universal and overcoming the barrier should be possible.

It was observed some time ago that DNA could be transferred from one *species* of bacterium to another. This represented a rather special system developed by bacteria for acquiring resistance to antibiotics. In evolutionary terms the development of this system meant survival. Antibiotic resistance frequently develops in bacterial species and is a hazard of the widespread use of antibiotics. Gradually the resistant strain becomes the predominant one and the antibiotic becomes less and less effective. This is usually because the resistant strain can produce an enzyme capable of inactivating the antibiotic—penicillin*ase* is an example of such an enzyme. The ability, i.e. the possession of the genetic information to make penicillinase and other antibiotic-inactivating enzymes can be transferred amongst bacterial species. Originally, it was found that *Salmonella* could acquire antibiotic resistance from *E. coli*, and it was soon shown that the information for producing the enzyme responsible was contained in a separate piece of DNA distinct from the *E. coli* chromosome. This extrachromosomal 'resistance' or 'R-factor' was a small circular piece of DNA double helix that was transferred from one cell type to another via a narrow 'pilus' that formed between two cells. R-factors belong to a class of extrachromosomal DNA molecules found in prokaryotes, called 'plasmids'. These are always relatively small DNA molecules (just a few genes) and are capable of self-replication in the host cell. They can be separated from the larger chromosomal DNA by centrifugation.

It became clear that if it were possible to 'manufacture' a plasmid containing a desired gene, it would be feasible to introduce it into an *E. coli* cell. There it would then self-replicate and potentially could be expressed. Introducing the plasmid into the *E. coli* cells is achieved by mixing plasmids and cells in the presence of calcium ions which make the cells permeable to DNA. This process is akin to bacterial transformation. (There are, in fact, alternative methods using viruses as 'vectors'.)

So, the question is how to 'manufacture' a gene and incorporate it into a plasmid. This is a problem of enzymology: living organisms manipulate DNA with enzymes during replication, transcription, fertilization, and so on. A knowledge of all the enzymes involved makes it possible to do almost anything with a DNA molecule: but what about manufacturing a gene? This is feasible with difficulty. Sequences of DNA 75–100 or more bases in length have been made, but these will only code for short polypeptides. It is much more realistic to try to 'extract and purify' genes, and one of the ways in which this may be done is as follows. Suppose we wanted to obtain the gene for insulin: the mRNA for insulin could be extracted from pancreatic β-cells, and if this is then treated, in the presence of the 4 DNA bases as triphosphates, with a virus enzyme called 'reverse transcriptase', a copy DNA, or cDNA, complementary in sequence to the RNA is produced. This is an unusual direction in which to go—usually the reverse applies, i.e. DNA → RNA—but 'reverse transcriptase' functions in certain viral systems which have RNA instead of

DNA as their genetic material, and only make DNA when they infect a host cell. Having obtained the cDNA, it can be treated with another enzyme, DNA polymerase, to produce double-helical DNA which represents the gene for insulin (Figure 6.4).

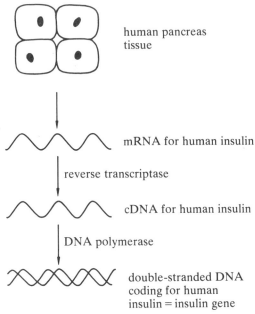

Figure 6.4 Isolation of the gene for human insulin

This gene now has to be inserted into a plasmid, and here again modern biochemical research has revealed a whole armoury of enzymes which make this a relatively easy task. Enzymes are known which make specific cleavages in a DNA double helix. These enzymes are obtained from bacteria and are called 'restriction endonucleases' or simply 'restriction enzymes', and over one hundred different ones are now known. Typically, a restriction enzyme recognizes a short, specific base sequence of about half a dozen bases in DNA and hydrolyzes the backbone at two points. Often these are not in the same place in both chains, so that two overlapping, complementary portions of DNA result. Because these are complementary they are called 'sticky ends' (Figure 6.5). The trick of inserting a gene into a plasmid is to add appropriate 'sticky ends' to the gene and then mix it with the restriction enzyme-treated plasmid (Figure 6.6). Finally, further enzymes are known that act as 'repair' enzymes. These 'mend' nicks in DNA and hence their action is to produce an intact double helix from the plasmid-plus-gene. The repair enzymes presumably have the function of looking after the maintenance of DNA in the cell, but they also have a role to play during replication.

Thus an intact plasmid can be produced that contains an inserted piece of DNA which could in principle be a gene for *any* protein. Such a plasmid is called a 'recombinant plasmid' as it has recombined with a piece of 'foreign' DNA. There are other ways of doing this and no doubt the next few years will see a great expansion of the technology for manipulations of this kind.

Restriction endonuclease 'Hind III' recognizes the sequence of bases in DNA shown below and makes 'nicks' in the sugar—phosphate backbone at the places shown by the arrows

$$\downarrow$$

5'————CCAAGCTTGG————3'

3'————GGTTCGAACC————5'

$$\uparrow$$

You may notice that the site recognized has a certain symmetry about it: this is typical of the sites recognized by these enzymes. Different enzymes recognize different sequences. Typically a single plasmid will have a single sequence like the one shown above, so that treatment with a given restriction enzyme will make a single 'nick'. The ends produced are complementary and overlap. They may be pulled apart under conditions that prevent hydrogen bonds forming. Conversely, in conditions that *favour* hydrogen bond formation (called 'annealing') such sequences will tend to match up and form hydrogen bonds. Thus any piece of DNA having the sequence shown below on its 'sticky end'

5'——CCA   will anneal with   AGCTTGG——3'

3'——GGTTCGA   ACC——5'

Figure 6.5 Action of a restriction enzyme on a piece of DNA

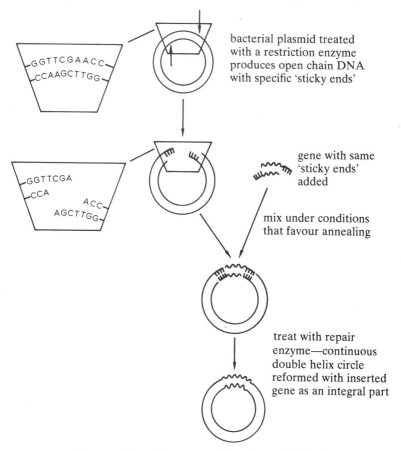

bacterial plasmid treated with a restriction enzyme produces open chain DNA with specific 'sticky ends'

GGTTCGAACC
CCAAGCTTGG

gene with same 'sticky ends' added

mix under conditions that favour annealing

GGTTCGA
CCA
ACC
AGCTTGG

treat with repair enzyme—continuous double helix circle reformed with inserted gene as an integral part

Figure 6.6 Inserting a gene into a bacterial plasmid

In the above we have only sketched out the possibilities: the technology, in detail, is complicated and, in addition, there are possible dangers (Box 6.5). It is hoped that enough detail has been given to enable the reader to understand that these things are now feasible. Insulin has already been produced in this way, and interferon, a protein that appears to protect cells against viral infection and may also have anticancer activity, is well on the way to being produced. Many of the peptide hormones are likely to be produced in this way in the future—grown in large cultures of bacteria, rather than extracted tediously from the endocrine glands of animals.

Once a bacterium has been 'tailored' to produce a human protein, the culture is, of course, immortal and can be grown on the 1 000 litre scale. One advantage of producing human proteins in this way is that when it is intended to use them in humans, they will provoke no antibody response. Diabetics may produce antibodies to injected *pig* insulin: there is no way at present of obtaining human insulin.

Box 6.5 Biohazards of genetic engineering

It was clear from the start that the construction of new genetic combinations might be hazardous. For instance, if the gene responsible for the production of cholera toxin were to be inserted into the gut bacterium *E. coli* in the laboratory, and then this *E. coli* 'escaped' from the laboratory and infected a human gut once more. Even worse if a gene from a cancer-producing virus (see p. 259) were to be inserted similarly. Cloning might result in the formation of completely new sequences of DNA which, if expressed, might have unpredictable, but disastrous consequences for the public at large.

Because of all these 'unknowns', a group of researchers brought together in 1974 in the USA expressed their 'concern about the possible unfortunate consequences of indiscriminate application' of recombinant DNA techniques. Possible types of experiment were classified and graded, and laboratories were designated with the idea of *physical containment* of the possible biohazards. The aim was to use stringent safety precautions so that there was no chance of modified organisms 'escaping' from the laboratory, and that the investigators themselves be protected by masks and gloves and the use of sterile boxes, etc. Some experiments, at least in some countries, were banned completely because of potential, unknown hazards.

The debate continued, and to some extent is still going on. Experts in law and ethics also became involved in the debate, and public opinion was tested. It cannot be said that the problem is resolved yet. While the scientists originally involved in these projects clearly behaved in a highly responsible way, other scientists have since taken the view that the dangers have been grossly overstated. If the dangers now seem less, it is partly due to some overstatement originally (we can now say with hindsight), but also due to new knowledge of genes and their behaviour in organisms, which has lessened the concentration of hypothetical hazards. In addition, a major development has been the use of *biological containment*. This means using strains of bacteria (*E. coli* for example) that have special nutritional requirements. Such organisms will only grow on special nutrient media in the laboratory, and strains can be selected which will not grow in the human intestine because these nutrients are not present. There is little chance of such an organism surviving and becoming a hazard even if it did manage to escape from the laboratory.

In the longer term, it may be possible to learn how to transfer genes successfully between eukaryotic cells or from prokaryotes to eukaryotes. If, for example, the genes for the enzymes for nitrogen fixation (p. 159) could be inserted into plant cells, there would be major advantages for agriculture.

Certainly the possibilities appear to be enormous and the technical background and biochemical knowledge are increasing apace. One has only to look at the newspapers or the popular science journals such as *New Scientist* or *Scientific American*—almost any issue at present—to see the impact of this technology on the way biochemists and molecular biologists are thinking.

6.6 Biochemistry and medicine

Perhaps the most 'humanitarian' aspect of biochemistry shows itself in the field of medicine. During their first years of training, medical students divide their time approximately equally between the study of the structure of the human body (anatomy), its biological functioning (physiology) and its chemistry (biochemistry). Frequently, disorders of body chemistry manifest themselves as diseases: this is why doctors need to have a detailed knowledge of biochemistry. In medical treatment, chemical substances (drugs) are used that interfere with the chemistry of the body in order to achieve certain effects. Other substances (antibiotics) are used to kill invading organisms selectively. Finally, the lack of certain natural substances such as the hormones, thyroxine and insulin is remedied by replacement therapy.

The role of the biochemist is to help the medical practitioner to (1) *diagnose* what is wrong, incidentally causing as little distress to the patient as possible in so doing, and (2) *devise* possible remedies. We would not wish to pretend that biochemistry alone is responsible for the changes that have occurred in medicine since the sulphonamides and the antibiotics were discovered in 1930–1940: pharmacology, bacteriology, chemistry and virology and many other disciplines have played major roles too. We *would* like to make the point that an understanding of the chemical reasons for disease states forms an excellent basis for attempts to devise cures.

Diagnosis

We look first at how the doctor decides what is wrong with the patient. The patient knows something is wrong because he feels pain or discomfort. The doctor uses his eyes, ears and hands, and his initial impression, taken together with his vast experience, to produce a diagnosis, i.e. 'this patient has a broken leg/jaundice/a virus infection'. Almost always, however, the doctor will also ask for blood and/or urine tests on the patient. The examination of body fluids is very important in arriving at a sound diagnosis of the disease (Box 6.6).

Box 6.6 Chemical pathology and diagnosis

The Chemical Pathology Laboratory of a hospital performs chemical and biochemical tests on samples taken from patients. A 'test' usually means determining the concentration of a specified substance. Most frequently samples of blood and urine are tested.

These days most of the determinations are done with automated equipment. A whole battery of tests is available and the doctor merely ticks a form requesting the values he is interested in. Some tests are done by straightforward chemical methods. In other tests the *levels* of enzymes in the serum or plasma are measured. The logic here is that if the cells of a tissue are damaged because of disease, enzymes characteristic of that tissue will be released into the blood. In bone cancers, for example, the blood level of the enzyme alkaline phosphatase increases because there is increased osteoblastic activity.

If the doctor knows the *normal range* of values for a particular blood component for healthy individuals, he can quickly see if his patient's blood has an abnormal concentration.

Although diagnosis requires great knowledge and experience coupled with an understanding of the many factors involved, the following description of the tests shown here should illustrate how they contribute to enabling the doctor to reach a conclusion.

Ions—sodium, potassium, chloride, bicarbonate, calcium, phosphate. All will be abnormal in kidney diseases as well as when the water balance of the body has changed, such as following severe diarrhoea and vomiting. Deficiency or excess of certain hormones and vitamins also affects the levels of some ions. For example, calcium and phosphate levels will be found to be abnormal in parathyroid gland diseases and when there is vitamin D deficiency or excess. They will also be disturbed in bone disorders such as bone cancer and osteomalacia. The bicarbonate level is an index of the acid-base balance of the body.

Urea and creatinine are nitrogenous waste products and their levels are used as an indicator of renal function.

Protein and albumin levels in the blood change in malnutrition and malabsorption, in liver cirrhosis, and in infection and inflammation.

Bilirubin is a breakdown product of haemoglobin. A raised level therefore may indicate an increased rate of destruction of red blood cells (haemolytic, jaundice) or it may indicate cholestasis (obstructive jaundice, e.g. from gall-stones).

Alkaline phosphatase levels are indicators of hepatitis, cirrhosis, bile duct obstruction, bone cancers, oesteomalacia, rickets and ulcerative colitis.

GOT means glutamate-oxaloacetate transaminase, an enzyme of amino acid metabolism. The heart, liver, red cells and muscle contain high levels of this enzyme. Destruction of cells of these tissues caused by disease will increase the blood concentration of GOT. Examples of such diseases are hepatitis and myocardial infarction (heart attack).

The hospital biochemist has a background knowledge of what substances are normally found in blood and urine and also knows the 'normal range' of values expected for healthy individuals. The general idea is that in disease, abnormal substances, or normal substances in abnormal concentrations, may be present in the body fluids. The biochemist's role is to devise tests to identify and quantify these.

When a diagnosis has been made and drug or other treatment started, it is advantageous to continue to analyze blood samples to find out if the treatment is successful, is having no effect, or actually making the condition worse. Very often early warning signs of the side-effects of drugs may be picked up on the basis of an abnormal blood 'profile'. This sort of analysis also contributes to the discovery and development of new chemicals for use as drugs.

Tests on body fluids may also give an indication of dangers to come. For example, the concentration of cholesterol in the blood may signify the likelihood of heart disease in the future.

Chemotherapy

The word 'chemotherapy' means 'treatment by means of chemicals'. In earlier times, herbalists prescribed plant products to cure ills, and nowadays many people seem to believe, wrongly, that there is 'a pill for every ill'.

Louis Pasteur was the man responsible for the 'germ theory of disease'—the idea that diseases were caused by the 'microbes' we now know to be bacteria, viruses, etc. Not all diseases are in fact caused by micro-organisms, but in Pasteur's day the majority of people dying from disease died of microbial infections—plague, cholera, diphtheria, typhus, TB, influenza, and so on.

Paul Ehrlich noticed that certain dyes stained bacteria, and hoped to find a dye which would seek out and kill a bacterial cell inside a human body whilst not affecting the host cells. By 1907 Ehrlich had discovered 'trypan red' which would stain and kill the trypanosomes responsible for sleeping sickness (transmitted by the tsetse fly) without killing the patient. Later, he discovered 'Salvarsan', an arsenic compound containing the grouping $-As{=}As-$ which kills the spirochaete that causes syphilis. Salvarsan was the first truly synthetic drug as opposed to naturally-occurring plant remedies such as quinine.

Much later (1932), a new red dye called 'Prontosil' was used against streptococcal blood poisoning, but it was found, curiously, that this drug was inactive against the bacteria in the test tube. 'Prontosil' is broken down in the body to give a compound, sulphanilamide, which is the compound actually responsible for killing the bacteria.

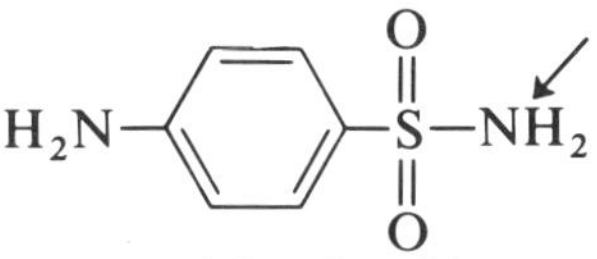

sulphanilamide

Sulphanilamide was the first of the 'wonder drugs'. By substituting different chemical groupings on the nitrogen atom attached to the sulphur atom, a range of drugs was produced which were effective against different bacteria and had fewer side-effects.

Sulpha drugs, of which sulphonamide itself is the parent compound, appear to act as competitive inhibitors (p. 72). Their structure is very similar to that of p-aminobenzoic acid:

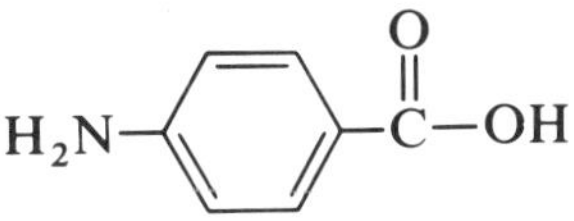

which is required by bacterial, but not human cells. Bacteria use p-aminobenzoic acid to manufacture a vitamin called folic acid. The enzyme responsible for the conversion of p-aminobenzoic acid to folic acid is competitively inhibited by sulphonamide and its derivatives. Because the bacterium, in the presence of sulphonamide, cannot produce folic acid, it dies. Human cells, in contrast, are unaffected—they do not make their own folic acid, but require it to be supplied (from the diet), and have no enzymes that can be inhibited by sulphonamide.

Soon after the sulphonamide drugs had been discovered, the first of the antibiotics, penicillin, appeared on the scene. Many other antibiotics followed. The mode of action of penicillin is similar to that of sulphonamide, in that it is a competitive inhibitor, in this case of enzymes involved in bacterial cell wall synthesis. Other antibiotics act by interfering with certain steps in protein biosynthesis in bacterial, i.e. prokaryotic, cells, but not human, i.e. eukaryotic, cells because the processes in the two types of cell differ in detail.

The consequence of the discovery and widespread use of antibiotics, and the use of immunization, is that in the developed regions of the world at least, men and women can on average be expected to live for 70 years, whereas at the time of Pasteur the average was less than 40 years. Humans now live long enough to be killed by heart disease, cancer, and other diseases, rather than being killed at a younger age by bacterial infections!

Viruses

What are the possibilities of producing chemicals or antibiotics for use against virus-induced diseases? Viruses are much more difficult to attack once they are inside the body. Bacteria and other parasitic organisms have their own metabolism and enzymes, and potentially should be susceptible to attack in a variety of ways. Viruses, on the other hand, simply 'commandeer' the host's cellular machinery, and so to attack the virus would be to attack the host equally.

In spite of this, there are some agents which *can* affect viruses and which have found a use in medicine. Some act extracellularly, attacking the free virus and others attack at the point where the virus is entering the cell: yet others act at the stage of synthesis or assembly of new virus particles. The main problem is to find some feature of the life cycle of the virus differing from that of the host cell, and this is not easy. One way of attacking virus-induced diseases is to use derivatives of nucleotide bases, such as fluorodeoxyuridine.

thymidine   fluorodeoxyuridine

These analogues are presumably incorporated into viral DNA, or inhibit enzymes that are making viral DNA, stopping virus synthesis, but it is not clear why the host cells are not affected equally. In general, antiviral chemotherapy is much less successful than antibacterial chemotherapy. The main reason why many virus-induced diseases such as poliomyelitis are under control is that there are programmes for inducing immunity in whole populations by vaccination. In a different way, yellow fever and typhus, both viral diseases, have been controlled by the use of DDT to kill the vectors, mosquitoes and body-lice, respectively. There has been much interest recently in interferon, a naturally-occurring antiviral agent.

Immunology

It is appropriate at this point to say something about immunology and the immune system. The science of immunology goes back to the time of Pasteur or even before, but only comparatively recently has it been possible to understand the chemistry of the various immune reactions. Over the last 10–20 years the complicated structure of the antibody molecule has painstakingly been worked out, and in the last 2–3 years it has become possible to understand how antibody molecules are synthesized. It is remarkable that the body has the potential for making somewhere between several thousand and a million types of antibody molecule, but only makes some hundreds of these at any given time. Each cell of the mammalian body certainly does not have a million genes coding for a million different antibody molecules. Rather the genes exist as short sections of DNA which can be combined in all sorts of ways to obtain, eventually, antibody molecules with different shapes that are complementary to antigens. A mechanism exists, at present poorly understood, for selecting those cells which make the most appropriate types of antibody molecule at a given time.

Another area where knowledge is increasing rapidly is that concerning the various factors involved in the rejection of 'foreign' tissues. An understanding of the complex biochemical processes that occur when a graft is rejected offers hope that transplants (kidneys, heart, skin) will survive longer (see below).

The combination of biochemistry and immunology is a formidable one, and we are likely to see major steps forward in understanding immunological phenomena, in developing new techniques and in treating disease in the near future.

Cancer

This account of biochemistry in relation to medicine would not be complete without reference to the group of diseases classified as 'cancers' (Box 6.7). In cancerous conditions certain cells of the body cease to respond to the normal controls on their growth. They multiply and crowd out, invade and destroy other tissues, and may eventually kill the individual. An enormous amount of biochemical research is aimed at trying to determine the underlying reasons why cells suddenly start to divide in this uncontrolled way. Much pharmacological research is aimed at evaluating agents—chemical and physical—that are potential cures for cancer.

It is not possible at present to say why, in biochemical terms, cell division suddenly goes out of control. We can recognize five broad types of causative factors: certain chemicals, physical agents, some viruses, 'premalignant states' and genetic factors. We will deal briefly with each of these in turn.

Box 6.7 Cancer: some definitions

Neoplasia. Normally, body tissues are continuously replaced to make good 'wear and tear'. After injury or disease this process of repair and regeneration goes on at an increased rate. In cancer, proliferation of cells occurs at a rate that exceeds that required for replacement or repair, and the new growth of tissue is spoken of as a *neoplasm* or a *tumour*. The growth of the tumour mass is uncontrolled and often extends without regard to normal boundaries, invading or crowding out adjacent tissues. Breaching of blood vessels by the growing tumour may result in transport of *neoplastic cells* far beyond the limits of the original tumour. When cells, carried in this way to other parts of the body, grow to form secondary tumours, the process is known as *metastasis*.

Benign or malignant. Tumours may be benign or malignant, though these two terms actually represent the two ends of a spectrum describing the rate of growth of a tumour. Some tumours never progress beyond a certain stage and do little harm: others grow widely and rapidly, resulting in death. The differences are summarized in the table:

	Benign	*Malignant*
Structure	Well differentiated, resembling tissue of origin. May possess a fibrous capsule	Tendency towards loss of structural and cellular differentiation. Unencapsulated, invades surrounding tissue
Rate of growth	Slow and may cease	Usually rapid
Clinical effects	Limited: mechanical pressure or hormonal	Usually lethal due to widespread destruction of tissues

Tumours are classified according to the tissue from which they have arisen. Thus a *fibrosarcoma* is a malignant tumour of fibrous connective tissue, and a *liposarcoma* is a malignant tumour of adipose tissue. *Leukaemias* arise from tissues that form the white blood cells and the blood may contain large numbers of white cells sometimes at a primitive stage of development.

Carcinogens. A carcinogen is a substance or a physical agency that causes cancer. The word *oncogenic* has a similar meaning: strictly oncogenic means 'tumour-producing' from the Greek *onkos*, a mass or a tumour. Viruses that cause tumours are called oncogenic viruses.

Chemical carcinogens.

It has been known for a long time that many human cancers result from exposure to toxic chemicals. Some are hydrocarbons found in tobacco smoke and tars, but the list includes many other types of organic chemical as well as some inorganic ones such as asbestos. It is suspected that many of these compounds are cancer-producing because they are *mutagenic*, i.e. they cause changes in the DNA. Dimethylnitrosamine, for example, can methylate the bases in DNA. One test for potentially carcinogenic substances currently in use is the Ames Test which determines whether the substance in question causes mutations in cultured bacterial cells. The argument is that if it causes mutations it may well also be carcinogenic.

Physical agents

Physical agents such as ionizing radiation can cause cancers. There is, for example, a relationship between excessive exposure to the ultraviolet component in sunlight during sunbathing and the incidence of certain types of skin cancer. Early workers with X-rays and radioactive materials also had a higher-than-normal incidence of cancer. These radiations are also mutagenic.

Viruses

It was shown in 1911 that certain connective tissue tumours in hens could be transmitted to healthy hens in a cell-free filtrate obtained by grinding up the tumour. Eventually RNA virus, avian sarcoma virus, was identified as the causative agent, and subsequently a number of other viruses were identified as being *oncogenic* or tumour-causing.

It is not clear whether malignant disease in man can be caused by a virus, although this has been claimed many times. The very fact that viruses cause disease in a wide variety of animals makes it highly likely that some human cancers are caused by viruses. The problem is that although virus particles may be present in human tumour tissue, it is difficult to establish whether they are there as *causative* agents or whether they are there as a *result* of the disease, by virtue of a diminished immunological response or a failure to produce interferon.

Premalignant states

'Premalignant states' refers to conditions that develop slowly as a result of continued irritation or prolonged exposure to small doses of carcinogens. Well-known examples are changes in the bronchial epithelia of heavy smokers, and ulcerative colitis.

Genetic factors

Genetic factors refers to a hereditary disposition to the development of neoplasms, but the significance and mechanism by which this happens is a mystery at present. For example, stomach cancers occur more frequently in people belonging to blood group A.

Theories of cancer

For many years it was believed that cancer resulted because a somatic mutation caused 'loss of control' over the rate of growth and proliferation of certain cells. More recently the association between viruses and cancer seemed to suggest a different cause for the onset of the disease. The true situation may be that both theories have some basis. In at least one form of cancer in mice and other animals caused by a virus, the polyoma virus, the viral DNA incorporates itself into the chromosomal DNA of the host cell. In this situation, instead of a mutation being introduced by a chemical or physical agent, it may be introduced as part of the viral DNA. Whatever the origin of the mutation, the cell becomes endowed with the capacity to multiply in an uncontrolled way, and to invade and destroy adjacent tissues.

Treatment

Many forms of cancer may be treated successfully with radiation (X-rays, radioactive emanations from radium-226 and other radioactive nuclides). The very radiation that can cause disease can also, if used appropriately, destroy the rapidly-dividing cell population that forms the tumour. Now, instead of producing a small change in the DNA, presumably a mutation, wholesale destruction of the DNA leads to death of the cell. Chemical substances can be used in the same way, and many of the chemotherapeutic agents used in cancer treatment, such as the mustards, are alkylating agents that can affect the DNA. Both radiation treatment and chemotherapy are potentially dangerous because of their non-specific action and unpleasant side-effects. They kill *all* cells not just the cancerous ones. Great care has to be taken that the doses used are sufficient to kill the rapidly-dividing cells of the tumour without too serious an effect on all the other cells of the body.

For some tumours, radiation has the advantage that it can be aimed at a particular, small area, without damaging other, healthy tissues. For other tumours, especially the so-called 'disseminated' ones such as those of the blood cells, e.g. leukaemia, the use of drugs that distribute themselves through the body is obviously more appropriate. Nitrogen mustards were first used in the treatment of human cancer, Hodgkin's disease, a type of leukaemia, in 1943, but fortunately since then very much less toxic alternatives have been produced. Mustards are extremely unpleasant compounds. They were original synthesized for use as nerve gases, and their side-effects are severe. Some of the modified forms are slower-acting, less toxic, and can be given by mouth instead of injection. Chlorambucil is one of the slowest acting and is effective in the treatment of leukaemias and certain other cancers.

$$CH_3-N \begin{cases} CH_2CH_2Cl \\ CH_2CH_2Cl \end{cases}$$

nitrogen mustard (mustine)

$$HO_2CCH_2CH_2CH_2-\langle\bigcirc\rangle-N \begin{cases} CH_2CH_2Cl \\ CH_2CH_2Cl \end{cases}$$

'carrier'   'war-head'

chlorambucil

Little is known of the distribution, metabolism and excretion of these compounds in man. With all of these chemicals there is the additional danger that they are likely to cause immunosuppression making the patient susceptible to infection. In addition to these, a number of other drugs may be used, with greater or lesser effectiveness, in the treatment of cancer.

Obviously, many more severe side-effects will be tolerated by a patient if the only alternative to treatment is almost certain death from the cancer. The pharmacologist and the biochemist have an important joint role in designing and evaluating more effective drugs having less unpleasant side-effects. At present, the prognosis for cure of cancer if diagnosed early is reasonably good. In addition to surgical removal of the tumour, preferably before metastases have developed, there is directed radiation and a whole host of chemotherapeutic agents to try. Many people are cured permanently. The biochemist contributes in a major way by providing a basis for understanding how and why the cellular machinery has gone wrong, and in suggesting how it may be put right.

Questions

1. (*a*) Briefly state the ways in which insects may be pests of man, crops and domestic animals.
   (*b*) For *one* defined situation of infestation by insects,
      (i) discuss how you would attempt to control pest numbers using management (cultural), chemical and biological methods,
      (ii) explain which control method you would expect to be most effective in (A.) the short term and (B.) the long term.

[JMB]

2. 'We envisage biotechnology—the application of biological organisms, systems or processes to manufacturing and service industries . . . will be of key importance to the world economy in the next century.' ('The Spinks Report', 1980.)
   Discuss the opportunities offered by 'biotechnology'.

[JMB]

3. Discuss the sources, effects and control of *four* marine pollutants (for example, oil, pesticides, radioactive wastes, organic wastes, inorganic sediment, heated water, heavy metals).

[JMB]

4. DDT when first used was extremely effective in killing insects such as mosquitoes. Explain the appearance of populations in which the mosquitoes are resistant to DDT.
   Why was the use of DDT stopped?

5. The following passage is taken from 'Microbes and Man' by John Postgate, reprinted by permission of Penguin Books Ltd. Read it and then answer the questions below.

   Modern methods of food processing and treatment sometimes, though far less often than food-faddists would have us believe, lead to products that are less wholesome than they might be. White bread, for example, is well known to lack several vitamins (E and many of the B group) present in wholemeal bread. So the practice has grown up of manufacturing nutrients of this kind in order to replace those lost in food processing, to enrich foods of limited nutritional value, and for use in medicine. Lysine is an amino acid derived from protein which human beings cannot synthesize: a certain amount must be provided in the diet, and it has been made industrially for the enrichment of bread. The process used is interesting because it makes use of two microbes in succession: one, a special strain of *Escherichia coli*, cannot make lysine because, as a result of mutation, it lacks a certain enzyme. It can only make an immediate precursor of lysine called diaminopimelic acid (DAP for short). If, then, this mutant is grown with only a little lysine (for it has to have some, or it will not grow at all), its lysine synthesizing system works normally up to the DAP stage, but stops there, with the result that relatively large amounts of DAP collect in the culture. Another organism, *Aerobacter aerogenes*, contains plenty of the enzyme necessary to convert the DAP to lysine, so, in the industrial process, this organism is grown, killed with toluene, and the extracted enzyme used to convert DAP, accumulated by the *E. coli*, to lysine.

   (*i*) Why does the *E. coli* mutant have to be given a little lysine?
   (*ii*) Why do humans need a supply of lysine in the diet?
   (*iii*) Why is only the enzyme from *A. aerogenes* used and not a growing culture of cells?
   (*iv*) Suggest why the *E. coli* mutant produces large amounts of DAP which collects in the culture fluid.

6. Water from Upper Klamath Lake is used to irrigate large areas of farmland in California, USA. From the fields, excess water passes into Tule Lake and from there to Lower Klamath Lake. In 1960, hundreds of dead water birds (gulls, herons, etc.) where found around Tule Lake and Lower Klamath Lake. The tissues of those and some aquatic organisms were analyzed and found to contain large quantities of insecticide residues as shown below:

Organism	Concentration of insecticide (p.p.m.)
Plankton	5
Fish	500
Birds	1 000

(*i*) What do you think was the origin of this pollution?

(*ii*) Suggest how the insecticide found its way into Tule Lake and Lower Klamath Lake.

(*iii*) Explain how the insecticide found its way, in such large concentrations, into the tissues of the birds.

(*iv*) Why did birds in Upper Klamath Lake show no ill-effects?

(*v*) Suggest one other type of pollution that could have arisen from the same source and explain how this would have affected the balance of nature in the lakes.

(*vi*) What do you understand by the term 'balance of nature'?   [Scottish Higher]

Further reading

Bonner, J. (1980), 'The World's Population and the World's Food Supply', *Oxford/Carolina Biology Reader*, **122**.

Bowman, J. C. (1974), 'Biology and the Food Industry', *Studies in Biology*, **46**. (Edward Arnold).

Boycott, J. A. (1971), 'Natural History of Infectious Diseases', *Studies in Biology*, **26**. (Edward Arnold).

Cohen, S. N. (July 1975), 'The Manipulation of Genes', *Scientific American*, **233**, 24.

Croce, C. M., and Koprowski, H. (February 1978), 'The Genetics of Human Cancer', *Scientific American*, **238**, 117.

Devoret, R. (August 1979), 'Bacterial Tests for Potential Carcinogens', *Scientific American*, **241**, 28.

Fisher, R. B., and Christie, G. A. (1975), '*A Dictionary of Drugs*'. Paladin 9 (Granada Publishing).

Gilbert, W., and Villa-Komaroff, L. (April 1980), 'Useful Proteins from Recombinant Bacteria', *Scientific American*, **242**, 68.

Inchley, C. J. (1981), 'Immunobiology', *Studies in Biology*, **128**. (Edward Arnold).

Meselson, M., and Robinson, J. P. (April 1980), 'Chemical Warfare and Chemical Disarmament', *Scientific American*, **242**, 34.

Nicolson, G. L. (March 1979), 'Cancer Metastasis', *Scientific American*, **240**, 50.

Noble, W. C., and Naidoo, J. (1979), 'Micro-organisms and Man', *Studies in Biology*, **111**. (Edward Arnold).

Novick, R. P. (December 1980), 'Plasmids', *Scientific American*, **243**, 76.

Parish, P. (1979), '*Medicines—A Guide for Everybody*'. (Penguin Books).

Postgate, J. (1975), '*Microbes and Man*'. (Pelican).

Taylor, I. J., and Senior, P. J. (1978), 'Single Cell Proteins: A New Source of Animal Feeds', *Endeavour*, **2**(1).

Index